→ INTRODUCING

EVOLUTION

DYLAN EVANS & HOWARD SELINA

This edition published in
the UK and the USA
in 2010 by Icon Books Ltd,
Omnibus Business Centre,
39–41 North Road, London N7 9DP
email: info@iconbooks.com
www.introducingbooks.com

Sold in the UK, Europe and Asia
by Faber & Faber Ltd,
Bloomsbury House,
74–77 Great Russell Street,
London WC1B 3DA or their agents

Distributed in South Africa
by Jonathan Ball,
Office B4, The District,
41 Sir Lowry Road,
Woodstock 7925

Distributed in Australia and
New Zealand by
Allen & Unwin Pty Ltd,
PO Box 8500,
83 Alexander Street,
Crows Nest, NSW 2065

Distributed in the USA
by Publishers Group West,
1700 Fourth Street,
Berkeley, CA 94710

Distributed in Canada
by Publishers Group Canada,
76 Stafford Street, Unit 300
Toronto, Ontario M6J 2S1

Previously published in the UK and
Australia in 2001 and 2005

ISBN: 978-184831-186-2

Originating editor: Richard Appignanesi

Printed and bound in the UK by Clays Ltd, Elcograf S.p.A.

The Central Idea in Biology

"Nothing in biology makes sense except in the light of evolution", said the great Russian geneticist, **Theodosius Dobzhansky** (1900–75). The theory of evolution is indeed the central idea in modern biology.

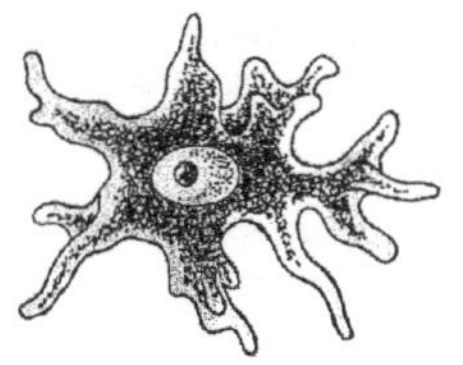

And yet, in August 1999, one hundred and forty years after Darwin published The Origin of Species, *the Kansas State Education Board* ***removed*** *the topic of evolution from the official curriculum.*

Biology students in Kansas would now be able to graduate without learning the most fundamental idea in their discipline! Why? What could motivate a State Education Board to deprive students of such an important piece of knowledge?

Fear and Loathing in Kansas

The members of Kansas State Education Board clearly disliked the theory of evolution. They are not alone. Ever since **Charles Darwin** (1809–1882) and **Alfred Russel Wallace** (1823–1913) proposed the theory at a meeting of the Linnaean Society in 1858, it has inspired much fear and loathing, and many attempts have been made to suppress it.

When the Bishop of Birmingham's wife heard about Darwin's theory in 1880, she said to her husband...

So what is it about Darwin's theory that upsets people so much?

Old Questions

Darwin's theory of evolution upsets people because it threatens all the old ideas about who we are and why we are here.

For thousands of years, human beings have wondered about the meaning of life.

The traditional answers, provided by many religions, have usually involved the idea of a God – or gods.

God made us, the story goes, and put us here for a special reason. All of these traditional answers see humans as exceptional creatures. Humans are not just animals. Unlike animals, humans have spirits or souls. Only humans have free will. Only humans can survive death.

Universal Acid

The theory of evolution threatens all these old ideas. It undermines the central claims of many religions. It seems to leave no room for God, or the soul, or life after death. Humans, it tells us, are just another kind of animal.

The American philosopher, **Daniel Dennett** (b. 1942), has described the theory of evolution as a kind of "universal acid".

Like universal acid, the theory of evolution eats through just about every traditional religious idea. This is why Dennett calls it "Darwin's dangerous idea".

An Idea in Two Parts

Darwin's dangerous idea comes in two parts: the theory of evolution, and the theory of natural selection. We will look at each of these two theories in turn, and then put them together. It is only when the two theories are put together that they become really dangerous.

We will begin with the theory of evolution.

What is Evolution?

The theory of evolution states that species can *change*. One species can give rise to another.

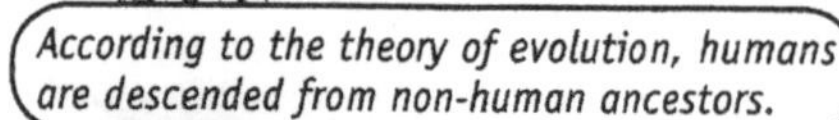

This might not seem like a big idea, but for thousands of years people in the West believed that species were fixed, unchanging entities. So it was a great shock to learn that this was not true.

The Fixity of Species

The idea that species are fixed and unchanging dates back to Aristotle (384–322 BC), the great Greek philosopher and scientist. He based his belief on the evidence of his own eyes.

So, not unreasonably, Aristotle concluded that one species could never give rise to another. Monkeys would always be monkeys. And humans must always have been human.

Independent Creation

For hundreds of years, Christian scholars accepted Aristotle's theory of the fixity of species. They believed that God had created each species independently at the beginning of time, and that each species then remained exactly the same up to the present.

"And now, from the clay of the ground, all the beasts that roam the earth and all that flies through the air were ready fashioned, and the Lord God brought them to Adam, to see what he would name them; the name Adam gave to each living creature is its name still." (Genesis 2:19).

Enough Time

Eventually, in the 18th century, some people began to suggest that species were not fixed. They realized that species might change gradually. A monkey could not have a human baby.

Given enough time. That was the crucial point. And that was what the Church disputed. Even if it were theoretically possible for one species to give rise to another by a series of many small steps, there simply had not been enough time for monkeys to evolve into humans, let alone for all life to be descended from a single ancestor. The world was simply not that old, according to the Church.

The Age of the Earth

When **Bishop Ussher** (1581–1656) added up all the figures in the Bible, he calculated that the world was created in 4004 BC. If the Bible was literally true, the earth could only be 6,000 years old. This was nowhere near enough time for evolution to take place.

Sir Charles Lyell (1797-1875)

Today, a mass of other scientific evidence has confirmed what those early geologists suspected. The world is almost a million times older than Bishop Ussher thought. Current estimates put the age of the earth at around 4.5 billion years. That is easily long enough for evolution to have taken place.

Old Bones

So evolution could have taken place. There has been more than enough time for all the currently existing species to have descended from a single common ancestor in a series of many small steps.

However, it is not enough to know that evolution *could have* happened. We want to know whether or not evolution *really did* happen.

In 1811, Mary Anning found the skeleton of an Ichthyosaurus, a marine reptile that lived between 5 and 120 million years ago, in cliffs near Lyme Regis in Dorset.

The 21 foot long fossil found by Mary Anning.

For thousands of years, people have come across old bones in the rocks. Some of these bones resemble those from animals we see around us, but others clearly come from animals unlike anything alive today.

Dragons' Teeth

Huge fossil teeth, for example, have been found that could not come from any currently existing species of animal. Where do they come from?

Legends and myths arose to explain these strange discoveries.

What the Fossils tell Us

The existence of strange fossils does not provide conclusive evidence against the fixity of species in itself. You can accept the idea that some species have become extinct without believing that one species can change into another. However, when you compare the various fossils that have been found, it is immediately obvious that they form *patterns*.

Many fossils can be arranged in sequences, in which there is a more or less continuous series connecting an earlier fossil with a later one.

The only good explanation for this pattern is that all these fossil species are related to each other.

The Evidence from Carbon Dating

Furthermore, carbon dating shows that the sequence of fossils as reconstructed by their appearance is the same as their sequence in time. The fossil at the beginning of the chain is older than the next one, which is older than the next one, and so on. If all species had been created independently, we should not expect them to appear in the fossil record in any order at all, let alone the exact order of their physical similarity.

The pattern of fossils thus shows that evolution is not just a theoretical possibility. It is a concrete *fact*.

The Tree of Life

From fossil and other evidence, biologists have been able to reconstruct the broad history of life on earth. They now know, for example, that all life on earth is descended from a single ancestor that lived about 4 billion years ago. This must have been a very simple kind of organism; much simpler even than a single cell.

The first cells appeared around 3.5 billion years ago.

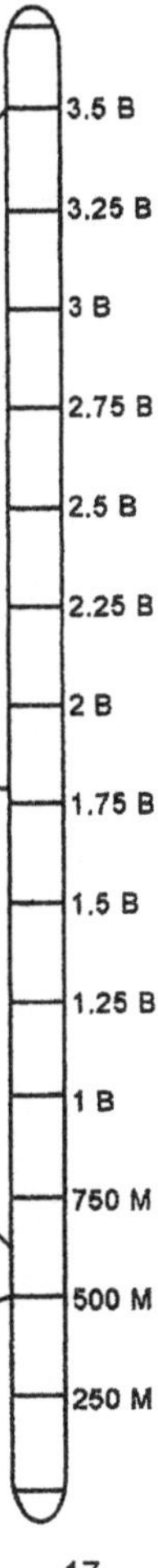

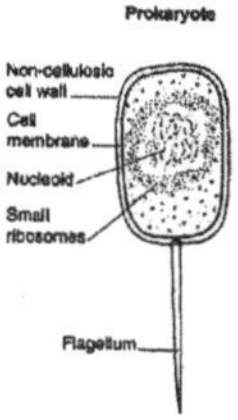

These were simple cells called prokaryotes, which have no nucleus.

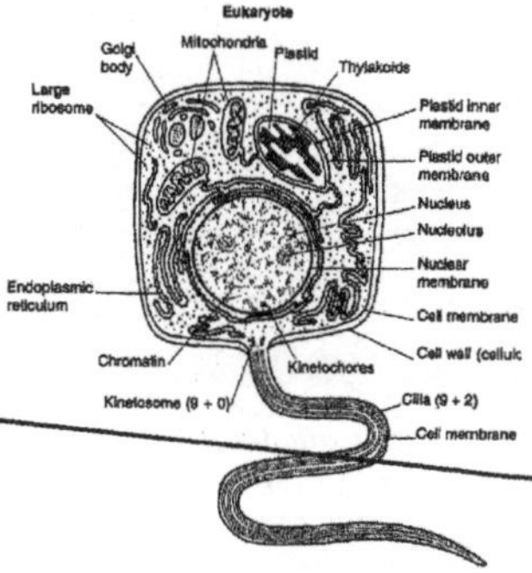

More complex cells called eukaryotes, which have a nucleus, appeared around 1.8 billion years ago.

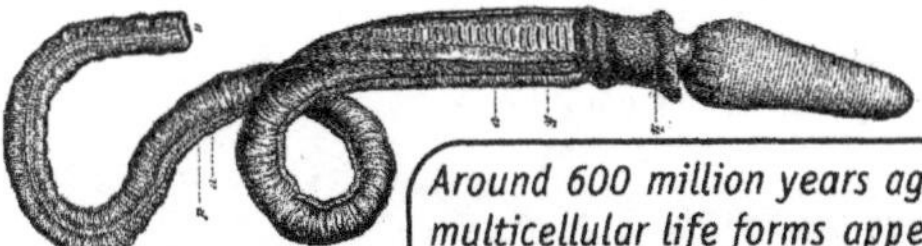

Around 600 million years ago, the first multicellular life forms appeared – jellyfish and worms.

Fish and plants evolved a hundred million years later.

The Conquest of the Land

Around 370 million years ago, animals began to colonize the land. Some were invertebrates, meaning that they had no backbone; these include the insects. Others were vertebrates, meaning that they had backbones.

Land-dwelling vertebrates include amphibians, reptiles, birds and mammals.

Four Legs Good

Vertebrates reveal their fishy ancestry in their skeletal structure – like the bony fish who first began to crawl out of the sea hundreds of millions of years ago, all vertebrates have four limb-like structures. The anatomical similarity of all these creatures is further evidence for evolution.

Amphibians and reptiles have four legs (though residual in snakes).

Whales are descended from cow-like ancestors that gave up their land-dwelling existence and went back into the sea. All that remains of their four limbs are tiny bony protuberances.

A Very Recent Species

Among the many kinds of mammal are the primates – monkeys and apes – which first appeared some 35 million years ago.

Modern humans (*Homo sapiens sapiens*) are all descended from this species of ape. Modern humans first appeared in Africa around 100,000 years ago. We are thus a very recent species. If the history of life on earth was compressed into a single year, humans would only make their appearance a few minutes before midnight on 31 December.

Living Fossils

The general outline of the tree of life is well known, but discovering the exact shape of the branches is quite hard. Fossils provide only part of the evidence. Other evidence lies in our genes.

We'll come back to genes in more detail later on. For the moment, just consider the following facts.

In other words, the common ancestor of humans and chimpanzees was more recent than the common ancestor of humans and bananas. All living things carry a record of the history of their descent in their genes. Genes are like living fossils.

"Scientific" Creationism

Despite the overwhelming evidence showing that life on earth has evolved gradually over millions of years, there are still many people who do not accept the theory of evolution. In the United States of America, for example, about a quarter of the population still believe in the literal truth of the creation story told in the book of Genesis.

Some fundamentalist Christians in the United States have even argued that **creationism** – the idea that God created all species in their current form a few thousand years ago – is a scientific theory on a par with the theory of evolution.

In 1987, the US Supreme Court declared that it was no more than thinly-veiled religion.

Missing Links

Without any good evidence to support their own position, creationists tend to fall back on trying to pick holes in evolutionary theory. One of their most common strategies is to point to gaps in the fossil record.

For some lineages of organism, fossils abound and we can arrange a continuous series of fossil skeletons to show how they evolved step by step. For other lineages, however, the series of fossils is not so continuous.

Fossilization is a precarious process, so it is not surprising that there are gaps in the fossil record. And there is so much *other* evidence for evolution that it would be unreasonable to worry about a few missing links.

How does Evolution Happen?

The evidence that evolution has occurred is now overwhelming. Creationism is no longer tenable. The theory of evolution is proven beyond all reasonable doubt.

This is where the second part of Darwin's dangerous idea comes in: the theory of **natural selection**.

Darwin's Contribution

It was the theory of natural selection – not the theory of evolution – that was Darwin's most original contribution to biology.

People had begun to suspect that species change long before Darwin.

So Simple, yet so Powerful

Natural selection is a very simple idea.

Thomas Henry Huxley

Yet natural selection is also a very powerful idea, for it can explain all the complex order we see around us in the biological world.

Three Conditions

Natural selection happens whenever the following three conditions are in place.

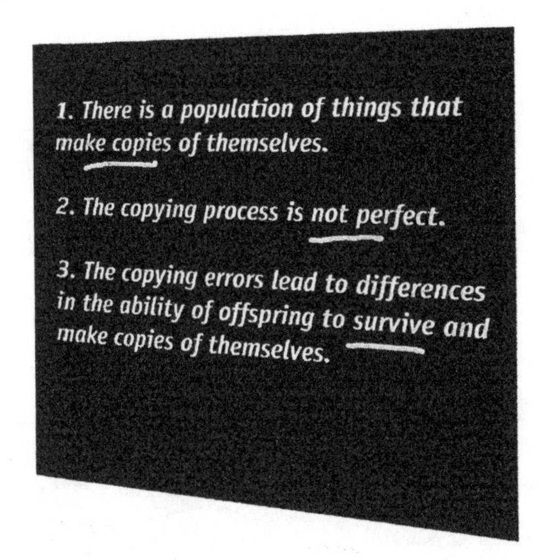

Note that these conditions don't just apply to animals and plants. They apply to *anything* that can copy itself.

Animal and Plant Copies

We will come back to computer viruses later, but for the moment let's stick with animals and plants. Animals and plants fulfil all three conditions for natural selection.

1. They make copies of themselves…

They have offspring that resemble the parents.

2. The copying process is not perfect…

The offspring differ in small ways from their parents.

3. The small differences between the offspring affect their chances of having offspring of their own.

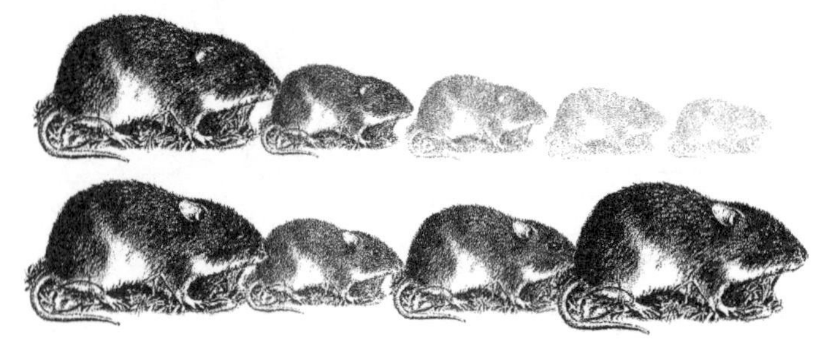

A Tale of Butterflies

To illustrate how the three conditions produce natural selection, here is a little story about a population of butterflies.

Butterflies make copies of themselves; that is, they have offspring that resemble the parents. So they meet the first condition. They meet the second and third conditions too, as will become clear in the story.

Once upon a time, there was a population of pale butterflies that lived in a wood in England.

When the butterflies sat on the pale tree branches, birds could not spot them very easily.

A Change in Environment

Then, one day, a big industrialist built a factory near the wood. The branches turned grey with the pollution.

Now the butterflies were no longer hidden when they sat on the branches.

Birds were able to spot the pale butterflies easily against a dark grey background.

The butterfly population began to decrease as more and more of them were eaten by the hungry birds. The butterflies continued to have offspring, but they did not live very long.

Multicoloured Butterflies

Most of the offspring were pale, like their parents. But reproduction is never perfect (**second condition**).

Now and again a butterfly was born with a different colour (of course, it was not a butterfly when it was born, but a grub which changed into a butterfly later).

Some of these colours made the butterfly even easier to spot against a grey background (**third condition**).

These butterflies were eaten while they were very young, before they could have offspring of their own.

One day, however, a grey butterfly was born. This butterfly was more camouflaged against the grey branches than the others, so the birds couldn't spot it so easily (this is another example of the **third condition**).

Most of these offspring inherited the grey colour. They too lived much longer than their pale cousins, and so had lots more offspring. After several more generations, almost every butterfly in this wood was grey. The population had evolved by *natural selection*.

Evolution by Natural Selection

Now we have looked at the theory of evolution and the theory of natural selection, it is time to put them together.

Put them together and you get the theory of *evolution by natural selection.* This states that species change into another species without any outside help.

Darwin's Dangerous Idea

The theory of *evolution by natural selection* is far more dangerous than the mere theory of *evolution*. We could, perhaps, accept that species evolve, and still believe in God. After all, might it not be the case that God supervises the evolutionary process?

It is not the idea of evolution itself that is a universal acid. It is the idea that evolution happens because of natural selection.

The Argument from Design

To see why the theory of evolution by natural selection makes the traditional idea of God superfluous, it is necessary to understand one of the most common arguments for the existence of God. This is the "argument from design".

William Paley (1743–1805) summed up this argument in his 1803 book, *Natural Theology*. When you look at a watch or any other complex machine, Paley said, you know that it must have been made by an intelligent creature.

Animals and Artefacts

Paley went on to observe that animals and plants also show the tell-tale signs of design. They are like machines, composed of exquisitely interlocking parts that all interact to help the organism survive.

The beak is made of a material much harder than tree bark, so it can be used to dig holes in trees, allowing the woodpecker to feed on insects that live under the bark and in the sap of the tree. Within the beak is a long tongue which has just the right shape for extracting insects from the hole in the bark. The woodpecker's muscles allow it to hammer away at the bark rapidly, while a stiff tail allows it to balance as it does so.

Adaptations

All these interlocking parts are well designed for a single purpose – to help the woodpecker get its favourite food. Such clever complex designs in organisms are called *adaptations*.

Paley assumed that this designer was God. In other words, just as the existence of a watch implies the existence of a watchmaker, so Paley thought that the existence of adaptations in animals, plants and other organisms implies the existence of a divine creator – God.

Another Explanation for Biological Design

The argument from design has impressed many people. Before Darwin it seemed that God was the *only* way to explain the complex design exhibited by living things. Thanks to Darwin, however, we now know that there is *another* way to explain the complex designs found in living things. This is natural selection.

A Single Step

Paley was right about one thing. Complex designs like the woodpeck-er's beak are highly unlikely to arise by chance in a single step.

The key phrase here is "*a single step*". A *simple* adaptation, like a change in colour, *can* arise by chance in a single step. This is what happened to the butterflies we saw earlier.

It would be ludicrous to suppose that at some point in the past, a beak-less parent gave birth to an offspring with a fully-formed beak. Darwin's theory does not appeal to such unlikely things.

Many Small Steps

Darwin realized that although a complex design like the woodpecker's beak is unlikely to arise by chance in a single step, it could easily arise by a series of many small steps. Each of these steps was a chance event in itself, requiring no designer, but natural selection would ensure that each step would be preserved and so the design would accumulate.

Perhaps the ancestor of the woodpecker was a bird with a very small beak.

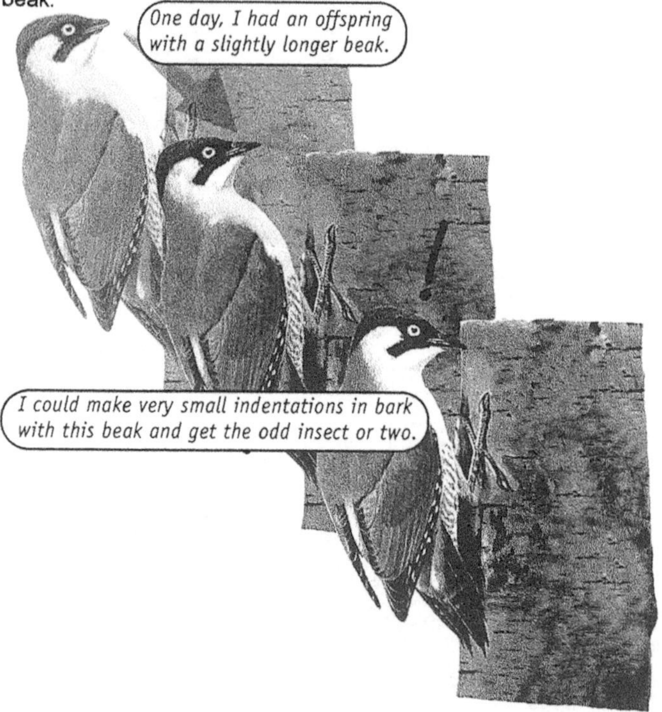

This gave it a very small advantage over its brothers and sisters, but very small advantages can sometimes be decisive in evolution. It passed its longer beak on to its offspring, and eventually the whole population had slightly longer beaks.

Start the Story Again

Up to now, the story of the woodpecker sounds very similar to the story of the butterflies. But, unlike the story of the butterflies, the story of the woodpecker does not end here, with a single-step change spreading through the population.

In the case of the woodpeckers, we can now start the story again.

Once again, this longer-beaked bird has a slight advantage over the other birds, and so has more offspring than the average bird. After many generations, the whole population has even longer beaks. And so on.

Cumulative Selection

We can continue this process repeatedly, adding small changes each time. The changes need not just involve longer beaks. The tongue might get longer too, for example. The main point here is that, unlike the butterflies, who changed colour in a single step, the woodpeckers evolved their complex beaks in a series of many small steps. Natural selection can be **cumulative** as well as single-step.

Each step must be a small improvement over the previous step, for otherwise the new design would not be copied more than the old one.

What Good is Half a Wing?

If evolution always proceeds by small steps, and each step must be an improvement, how could wings have evolved? After all, half a wing is not much good!

The answer is that, although half a wing might not be much good for *flying*, it might be good for *something else*. Something can evolve for one purpose, and then acquire another.

Then they learnt to glide, and finally to fly. Each step towards the evolution of the wing was an improvement in some way, but not all steps served the same purpose.

The Blind Watchmaker

Natural selection has no *foresight*. When feathers first evolved, this was not in order to prepare the way for the evolution of wings. Feathers just happened to evolve for one reason, and it so happened that this provided the material for the evolution of wings.

Natural selection designs wonderful things, but has no plans. For this reason, the British zoologist **Richard Dawkins** (b. 1941) has described it as a "blind watchmaker".

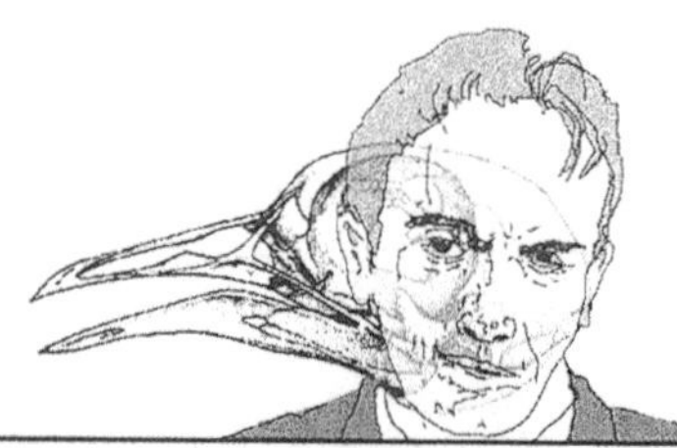

But because natural selection can be cumulative, the many small changes can eventually add up to a wonderful design. Cumulative natural selection is Darwin's answer to Paley's argument from design. Paley was right to demand an explanation for complex designs in nature. But he was wrong to assert that the *only* such explanation is God.

Occam's Razor

So, there are two possible explanations for complex designs. Either they were designed by God, or by cumulative natural selection. Which explanation should we prefer?

Whenever there is more than one possible explanation for something, there is a simple rule of thumb that we should use to decide which is the right explanation...

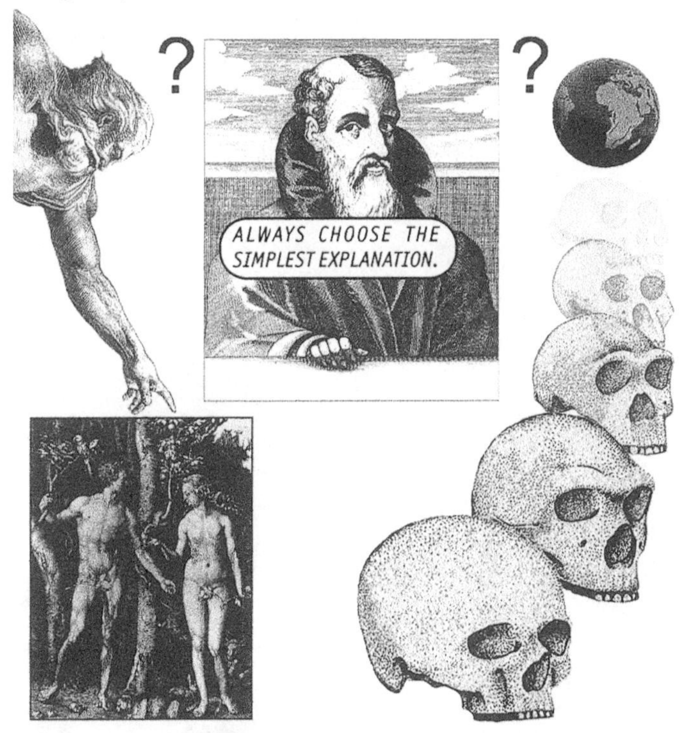

This rule is called "Occam's razor", after **William of Occam** (c. 1285–1347). Natural selection is a much simpler explanation than divine creation because it only requires us to believe in things that we already know about. All that natural selection requires are the three conditions described above. No supernatural being is needed.

No Need for that Hypothesis

Occam's razor tells us that we should always prefer simpler explanations. Natural selection is a simpler explanation than divine creation, since it only involves things we can already observe. Thus we should always prefer to explain biological design by natural selection rather than by divine creation.

This does not logically rule out the existence of God. But it does destroy one of the most powerful arguments for God. The reply of French mathematician **Pierre Simon de Laplace** (1749–1827) is now truer than ever.

So far, so good. This is a good *theoretical argument* for natural selection. But is there any *concrete evidence* for natural selection? Can we actually see it at work?

Natural Selection Observed

Yes, we can. The story of the butterflies is a case in point. This story is not pure fiction. Something very like it actually happened. The real story involves moths rather than butterflies, and the colours were slightly different, but the essential details are the same. The moths belong to a species called *Biston betularia* that is found in Great Britain. Before the industrial revolution, the moths of this species were always a light peppered colour.

The first dark moth was observed in 1848 near Manchester. A hundred years later, dark moths accounted for 90 per cent of the moth population in polluted areas. In unpolluted areas, the light form remained common. So, there are not just good theoretical arguments for natural selection – there is direct evidence too.

Looking in More Detail: Heredity

Now that we have seen what the main components of Darwin's dangerous idea are, we can take a look at the three conditions for natural selection in a bit more detail. (Before you go on, see if you can remember what these three conditions are. Then check page 27 to see if you are right.)

The first condition for natural selection is that there is a population of things that can copy themselves. Animals and plants produce copies of themselves when they have offspring. The offspring are "copies" because they resemble their parents. Monkeys have monkey babies, not human babies.

Even Darwin could not explain why offspring tended to resemble their parents. Now, however, we know why. We know something that Darwin didn't. We know about *genes*.

Genes

Genes are stretches of DNA. Each cell in our bodies contains lots of DNA. DNA is a complex molecule made up of four bases – adenine, cytosine, guanine and thymine (abbreviated to A, C, G, T). These bases are arranged in a long line. The sequence in which they are arranged is vitally important because different sequences specify different **proteins**.

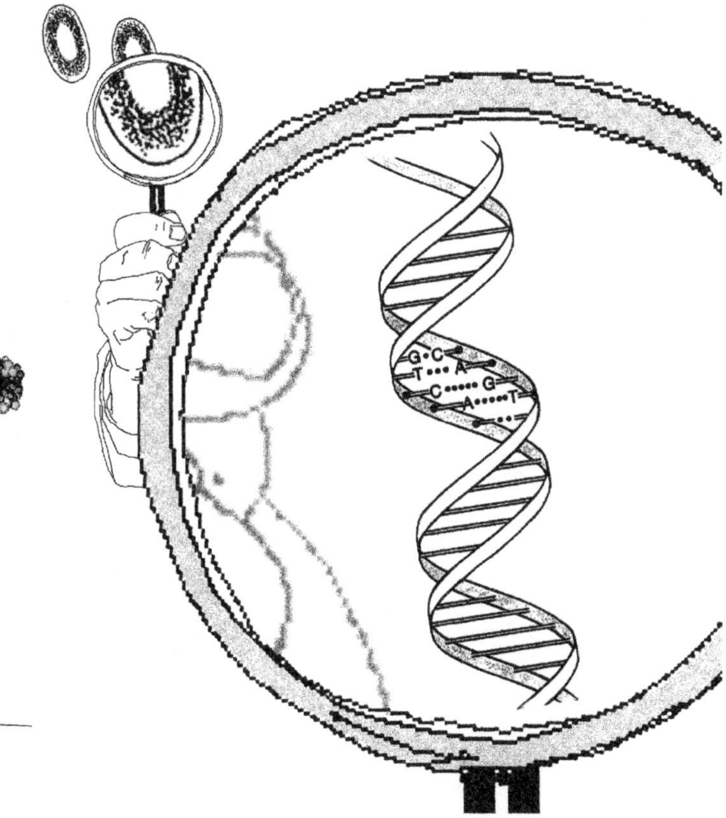

Proteins are the molecules from which animals and plants are made. They are the building blocks from which cells, and ultimately bodies, are constructed.

The Function of Proteins

Each gene is a particular sequence of bases and specifies a different kind of protein. Different species may have different kinds of body because they have different proteins, or because the proteins are arranged differently, or both.

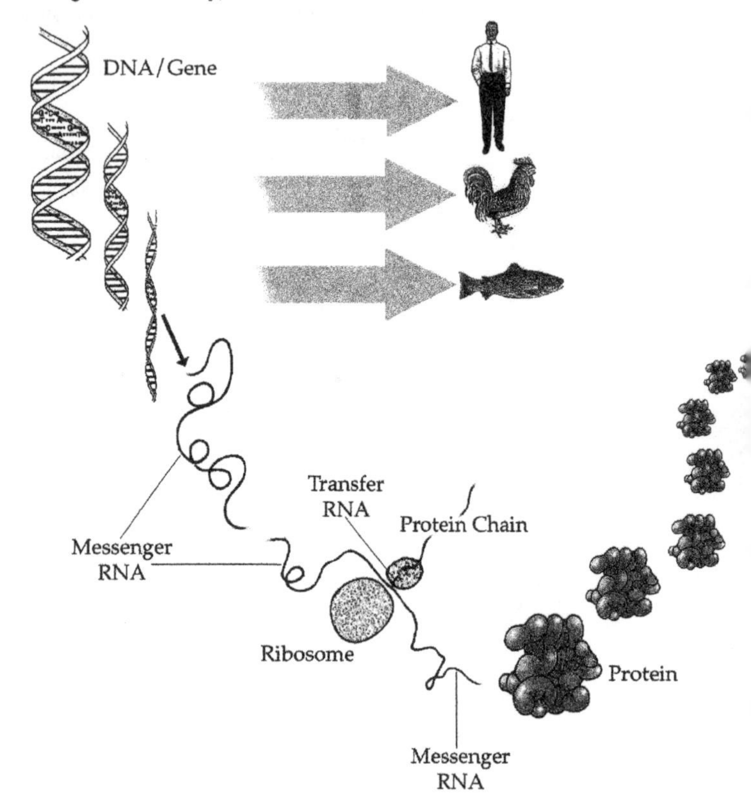

Encoded in the genes of an animal is the information that specifies both the kinds of protein in the animal and the way that these proteins are arranged.

Development

Every animal and plant starts life as a single cell. In order to grow into an adult, this cell must first divide in two (a process called *mitosis*).

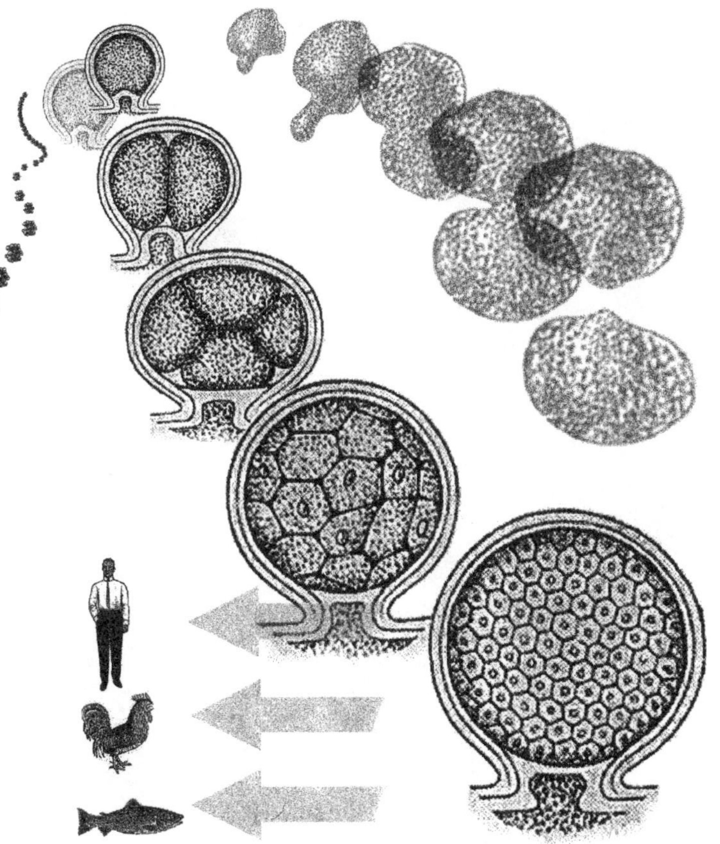

Each of these two cells must then divide again, and so on, until finally we have, say, an adult human being, or an adult monkey, each of which is composed of trillions of cells.

Growth and Specialized Cells

Growth does not just involve cell division. In addition to dividing and multiplying, the cells must specialize. The adult body of a human being is composed of many different types of cell: skin cells, brain cells, muscle cells, and so on.

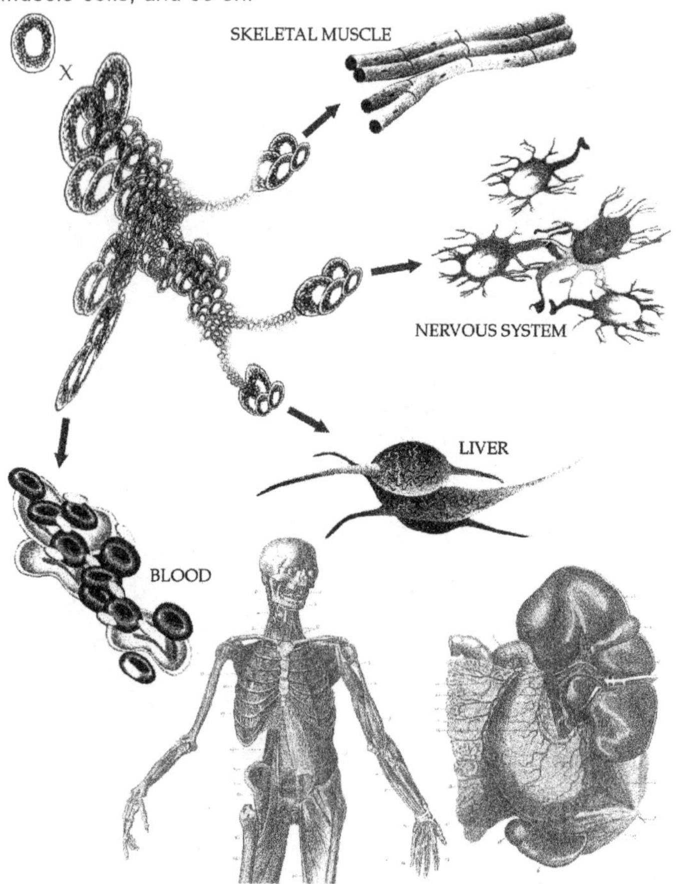

However, the initial cell is not specialized. It is a stem cell, with the potential to give rise to cells of any type.

Genes and Development

In order to grow into a fully-formed adult, the growing organism must make sure that the right kinds of cell are generated in the right order. Cells know how to do this by following the instructions encoded in their genes.

Of course, the difference is relative. Humans and monkeys are different, but not as different as humans and bananas. This is because humans and monkeys have lots of genes in common, while humans and bananas have fewer genes in common.

Genes and Environment

Take any two humans and compare them. They will resemble each other in many ways, because they have many genes in common, but they will not be perfectly identical.

Some of the differences between them will be due to the fact that they have different genes.

Despite the name, "identical" twins are never perfectly alike. The differences between identical twins cannot be due to genetic differences, since they have the same genes. All the differences between identical twins must therefore be due to differences in their environments.

Mutation or Incorrect Copying

The second condition for natural selection is that the copying process is not perfect. Some of the time, at least, the offspring must differ from the parents in some way.

It is easy to see that this happens in nature. Occasionally, a person is born with hair all over their body. Sometimes, a goat is born with two heads.

Some of these differences are down to the genes. Usually, our genes are faithfully copied from our parents. Half the genes come from the father and half from the mother. But occasionally, a gene is not copied correctly. The offspring begins life with a new gene not possessed by either parent. This new gene is called a **mutation**.

No Mutation, No Evolution

If mutation never happened, there could be no evolution and no natural selection. Every organism would be a perfect copy of its parents, and species would indeed be fixed. It is only because of the occasional copying error that species can change over time and adapt to new environments.

If all mutations were of this sort, natural selection could never occur. The third condition for natural selection states that the copying errors must affect the ability of offspring to make copies of themselves.

Random Drift

If all mutations were neutral, natural selection could not occur, but evolution could. Populations of animals could still change over time. Genes that have no effect on an organism's ability to survive and reproduce could still become more common or less common in the population by chance factors.

The existence of random drift shows that natural selection is not the only force driving evolution. The frequency of genes can change for reasons other than their effects on survival and reproduction.

Adaptation and Natural Selection

Evolutionary biologists argue about how important natural selection is in evolution. Some think that it is the most important force in the evolutionary process. Others think that random drift is at least as important.

Who is right? It all depends on what you mean by "the most important force". There are different ways of measuring evolutionary change, and, using some measures, random drift does come out as a very powerful force. One thing is clear though...

Not Everything is an Adaptation

Organisms are not just bundles of adaptations. In any organism, there are many features that are just by-products of other adaptations, or traits produced by random drift. For example, the white colour of our bones is not an adaptation – this colour has no particular purpose.

Adaptive Prediction

Sometimes, it is hard to know whether a trait is an adaptation designed by natural selection or a mere by-product. How can we find out which answer is right? There are several criteria that we can use to tell whether something is an adaptation or a by-product. The most useful is that of **adaptive prediction**.

Adaptive prediction works as follows. If there is a theory about how the trait in question helps its bearers to survive or reproduce, and this theory matches with the evidence, this favours the idea that the trait is an adaptation rather than a by-product.

If the trait has the features that are necessary to solve the problem, it is probably an adaptation.

Beneficial Mutations are Rare

Most mutations that are not neutral are harmful; they lead the organism to develop in ways that reduce its chances of surviving and reproducing. Such mutations are unlikely to get passed on to the next generation. Natural selection eliminates them.

Beneficial mutations are rare because when you have a complex design, such as an artefact or an organism, there are always many more ways of altering it that make it worse than ways of altering it that make it better.

Take a watch, for example, and imagine all the ways there are of altering it.

Because the watch is already well designed, most of these changes would make it worse. But the watch is not perfect, so there are probably one or two changes that would improve it.

Mutations are Random

Obviously, a human watchmaker does not proceed by making random changes. He does not usually take out springs or cogs blindly, and decide to alter them in any old way just to see what happens.

The mutations that appear from time to time in every species are completely random. Just because a mutation is beneficial does not make it more likely to occur.

We saw this in the story of the butterflies. When the branches became grey because of pollution, the best colour for the butterflies to be was no longer pale but grey.

Accumulation of Design

It may appear paradoxical that such a random process can lead to such complex designs. It is only the generation of mutations that is random, however. The conservation of beneficial mutations is *not* random. Obviously, only beneficial mutations are preserved.

Natural selection favours such rare beneficial mutations. The conservation of beneficial mutations is what allows the accumulation of design that we have already set out.

The Gene's-eye View

All this talk about whether or not mutations get passed on to the next generation can be used to look at evolution in a new way. As well as thinking of evolution in terms of changes in the *visible appearance* of a species, we can think of evolution in terms of changes in the frequency of the *underlying genes*.

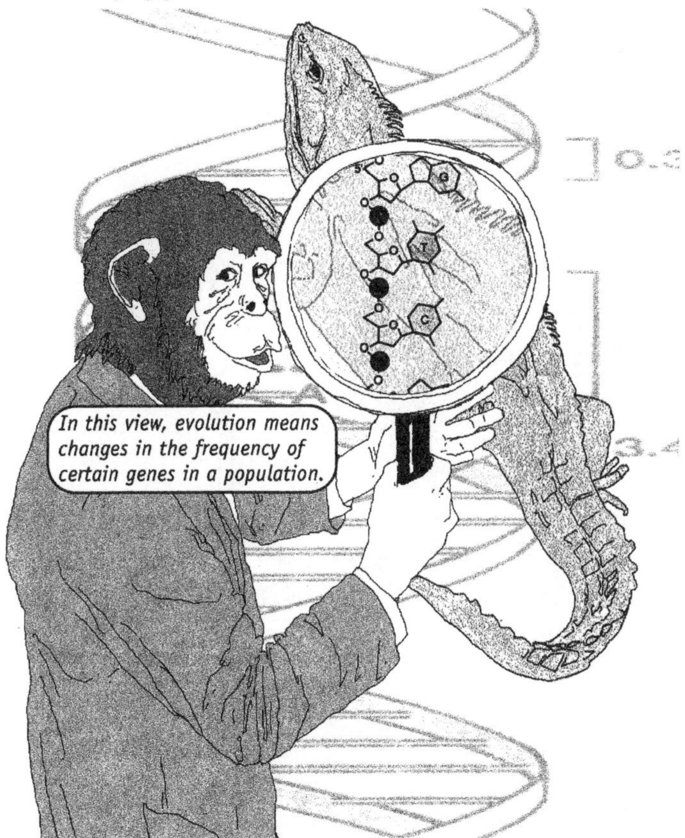

As one generation succeeds another, some genes become more common, while others gradually decrease in frequency and are eventually eliminated from the population.

The Selfish Gene

Richard Dawkins has also suggested that we can think of each gene "striving" to become more frequent in the population. This is only a metaphor, of course.

You want to spread more copies of yourself in the population. This is your only wish (you are very selfish). How can you achieve this objective?

Maybe you build sharper teeth. Or perhaps you build better muscles to make tigers run faster. Either way, the result would be to help tigers survive longer.

The gene would therefore achieve its "aim" of spreading more copies of itself through the population.

Altruism

It might seem as if the only way for a gene to spread through the population is by building adaptations, like teeth and muscles, that serve nobody except the one who possesses them.

But this is not, in fact, the case. Although genes are best thought of as "selfish" – that is, as striving only to make more copies of themselves – they can sometimes achieve their selfish objective by building things that benefit *other* organisms besides the one in which they sit.

Biological Altruism is Common in Nature

The biological definition of altruism is only concerned with effects, not motives. Unlike the normal psychological definition of altruism, in which motives are important, the biological definition does not worry about why the altruistic act was performed.

Altruism in the biological sense is very common in nature. Here is one readily observable example. Many social animals give warning cries when they spot a predator.

Parental Care

Another obvious example of altruism is parental care. Birds and mammals often engage in prolonged periods of care for their offspring.

When bees sting an animal, they die. They give up their lives to save their brethren.

The Puzzle of Altruism is Solved

The widespread existence of altruism in nature puzzled biologists for decades. From a selfish gene perspective, the puzzle becomes particularly obvious. After all, how can a gene achieve its *selfish* objective of becoming more frequent in the gene pool by making its bearer behave *altruistically*?

The answer is surprisingly simple. A gene which makes its bearer behave altruistically can spread through the population so long as *the altruism is directed specifically at other organisms who have the same gene*.

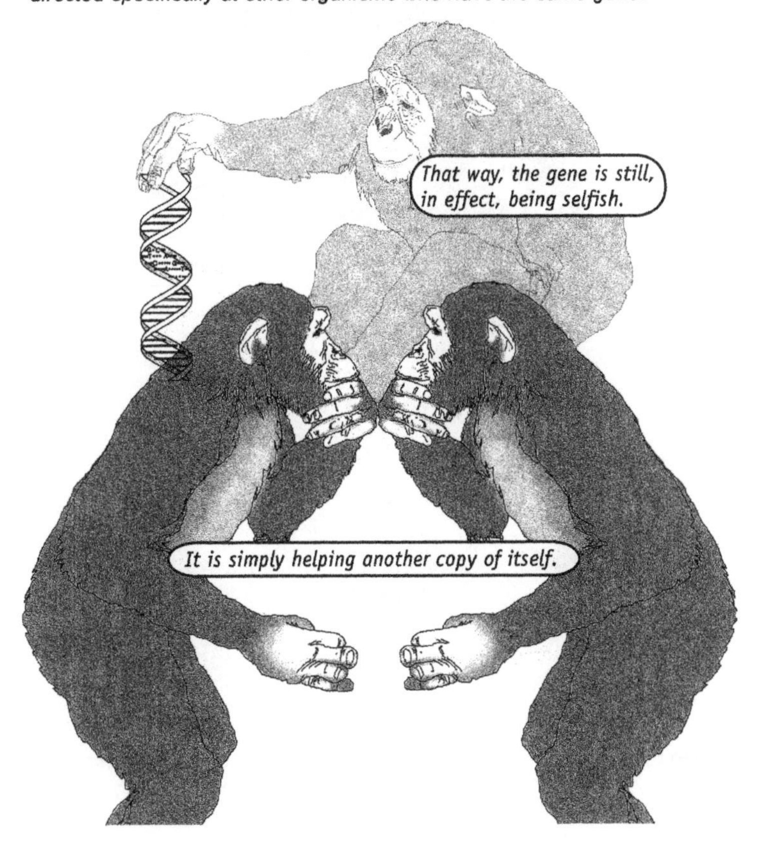

A Thought Experiment

A thought experiment will help to make this idea clearer. Imagine that you are a caveman and you find a tree with lots of fruit. Should you tell other cavemen about the tree or not?

Then they will have some food and survive. But you will have less than you would have done if you had kept quiet, so you have a higher chance of running out of food and dying.

Now look at things from the gene's point of view. Imagine you are a selfish gene sitting in the caveman's body. Your only wish is to make as many copies of yourself as possible. Should you make the caveman be selfish or altruistic?

According to the selfish gene theory, you should make the caveman be altruistic towards other cavemen who have copies of you inside them.

Their offspring will inherit the gene, and so the gene will spread through the population.

Two Conditions for Selfish Genery

For this selfish gene theory to work, two conditions are required...

1. Organisms must have some way of distinguishing between those who have the same genes as them and those who do not.

2. Genes must be

...*capable of influencing* behavioural *traits like selfishness and altruism.*

We can illustrate these two conditions with another thought experiment...

The Green Beard Effect

Suppose that there was a gene which made people grow green beards, and green beards were never found in people who lacked this gene. In this case, whenever two people with green beards met, they would be able to tell instantly that they both had the same "green beard" gene. The first condition would be fulfilled.

Now suppose that the green beard gene has another effect.

This hypothetical green beard gene would qualify as a selfish gene, even though it made people behave altruistically. It would, in effect, be helping other copies of itself.

The story of the green beard is, of course, fiction (Richard Dawkins made it up). We do not actually find green beards in nature. But we do find the two conditions illustrated by the story.

Do Genes Influence Behaviour?

Some people object to the idea that genes can influence behaviour. But whether or not we like it, the fact is that genes do affect the way we behave.

Hundreds of studies in dozens of species – from mice and rats to dogs and humans – have provided powerful evidence that genetic differences *between individuals* can cause those individuals to behave differently.

Genetic Determinism

Admitting that genes influence behaviour is not the same as "genetic determinism". It is not the case that every mouse that lacks this gene is more aggressive than every mouse that has it. The effect of genes on behaviour is statistical, not absolute. Nevertheless, this is enough for the selfish gene theory to work.

Can Organisms Tell When Others Have the Same Genes?

The second condition needed for the selfish gene explanation of altruism to work was that organisms must be able to distinguish those who have the same genes as them from those who don't.

In the story of the green beard gene, only organisms with green beards had this capacity. They were able to tell when other organisms had the green beard gene (like them) simply by looking.

Usually, organisms have to find some other way of telling whether or not another creature has many of the same genes.

Kin Selection

A simple way for organisms to find out whether or not others have the same genes as them is to find out whether or not they are closely related.

This observation led the British biologist **Bill Hamilton** (1936–2000) to propose that altruism will tend to evolve whenever organisms can direct it at those to whom they are closely related. This idea is known as the theory of **kin selection**.

Blood is Thicker Than Water

Since Hamilton put forward his theory of kin selection in the early 1960s, biologists have found lots of evidence that supports it. Almost all the examples of altruism described by biologists involve close relatives.

Parental care, by definition, occurs between animals that are closely related. But what about warning cries and self-sacrifice? These could conceivably happen between animals that are *not* closely related.

And the only animals that regularly engage in self-sacrifice to help others are social insects like bees who live in colonies in which everyone is *very* closely related.

Another Theory of Altruism

In 1973, the American biologist **Robert Trivers** (b. 1943) put forward another theory of how altruism can evolve. He argued that altruism could be favoured by natural selection if animals had a way of targeting their altruism specifically at those who were likely to reciprocate. This idea is known as the theory of "reciprocal altruism".

This is quite a tall order. How can you tell if someone will return favours or not? If you get it wrong, you will end up being taken for a ride.

Tit-for-Tat

In the early 1980s, an American political scientist called **Robert Axelrod** (b. 1943) showed that you can avoid being taken for a ride too often if you simply assume that others will behave as they have done in the past. If someone has returned your favours previously, they will probably do so again, and vice versa.

If they do reciprocate, then you should carry on doing them favours. If, one day, they stop reciprocating, then you should stop doing them any favours too. This strategy is known as "tit-for-tat".

Reciprocal Altruism is Rare

Tit-for-tat is a very simple strategy, but it still requires a lot of brainpower. Not only must you be able to recognize each individual you have met before, but you must be able to remember how each one has treated you in the past. Only animals with large brains are capable of this.

It is not surprising, then, that tit-for-tat is rare in the animal kingdom.

Since tit-for-tat is the most plausible basis for reciprocal altruism, it seems likely that reciprocal altruism is not widespread in nature. Most biological altruism, it seems, is due to kin selection, not reciprocity.

The Peacock's Tail

The existence of altruism among animals was not the only problem that baffled the first evolutionary biologists. The peacock's tail was another. What possible purpose could this ornament serve? How could natural selection favour such a thing, when it has a *negative* effect on survival chances? Big bright tails attract the attention of predators and make it harder to escape from them. Tails are also targets for parasites and are hard to keep clean.

So, a peacock with a small, drab tail would live much longer, but he wouldn't have any offspring. Genes for small, drab tails would soon die out.

Why do Peahens Prefer Big Tails?

Darwin's argument explains why peacocks have big, bright tails by reference to the idea that peahens prefer peacocks with such tails. Peahens do, in fact, have such a preference. When peacocks are given even bigger and brighter artificial tails, the peahens go wild!

But why do peahens have such a preference? According to some biologists, this preference has evolved because it helps peahens to mate with the healthiest males, and so have healthier offspring.

So the peacock's tail is an honest signal of fitness, and peahens who are attuned to this signal have a good way of sorting out the fit peacocks from the unfit ones.

As the example of the peacock's tail shows, natural selection is not just about survival – it's about reproduction too. If an organism lives to be 1,000 years old, but has no offspring, it might just as well have died at birth as far as its genes are concerned.

So, in sexual species, we find two kinds of adaptation.

Natural Selection and Sexual Selection

When an adaptation is designed for courtship purposes, like the peacock's tail, biologists say that it has arisen by *sexual selection*.

Female Choice and Male Competition

Sexual selection can lead to two main types of adaptation. One is the kind of adaptation "for courtship", like the peacock's tail. These adaptations help animals with them to win mates by appealing to the preferences of the other sex.

Winning mates is not only about getting chosen, however. It can also be about beating off the competition.

Both kinds of sexually selected adaptation are found mainly in males. Females do the choosing, so they don't need the costly ornaments and weapons that males need if they are to be chosen and to fight off their rivals.

The Runaway Process

Sexual selection can lead to a process known as "runaway". If, instead of preferring tails of a particular size, peahens simply prefer the biggest tail they can see, tails will get larger and larger with each generation.

Arms Races

Sexual selection is not the only cause of runaway evolution. A similar sort of thing can happen when two species are locked in an "arms race". For example, consider two species, one of which is the favourite food of the other. If the predator species has very sharp teeth, then the prey species might evolve thicker skin or a tougher shell.

The Antibiotics Arms Race

We can see an example of such an arms race in the growing resistance of some bacteria to antibiotics. Since the middle of the 20th century, doctors around the world have been using antibiotics in ever greater quantities.

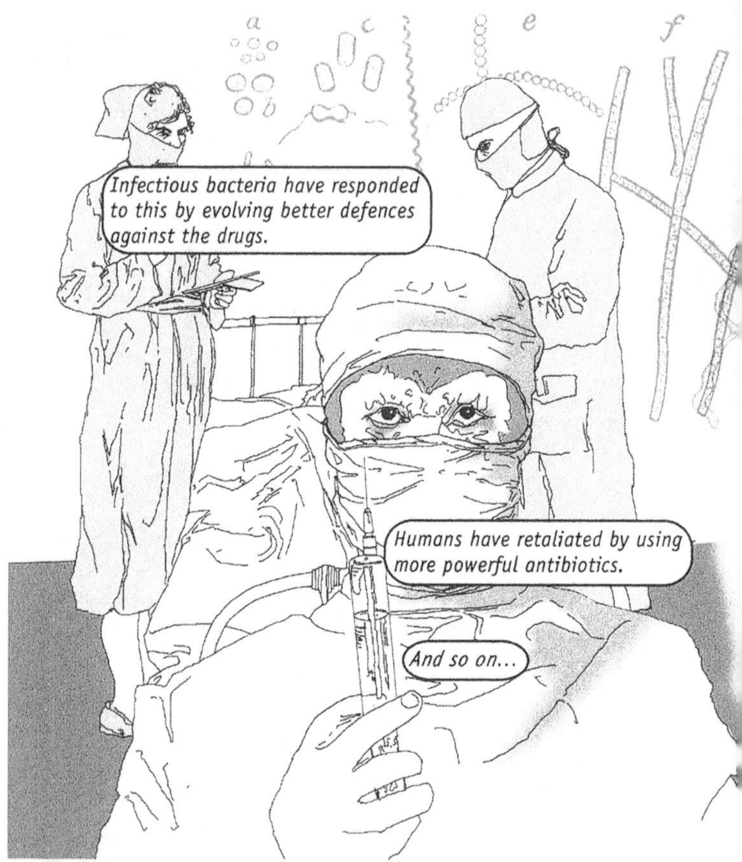

The result of this arms race is that some strains of disease are now resistant to all known antibiotics, and are therefore untreatable. Some see these "superbugs" as a major threat to humanity.

Co-evolution

Whenever two species evolve in tandem, with changes in one leading to changes in the other and vice versa, biologists call it "co-evolution". Arms races are one kind of co-evolution. But co-evolution need not involve conflict. It can involve co-operation too.

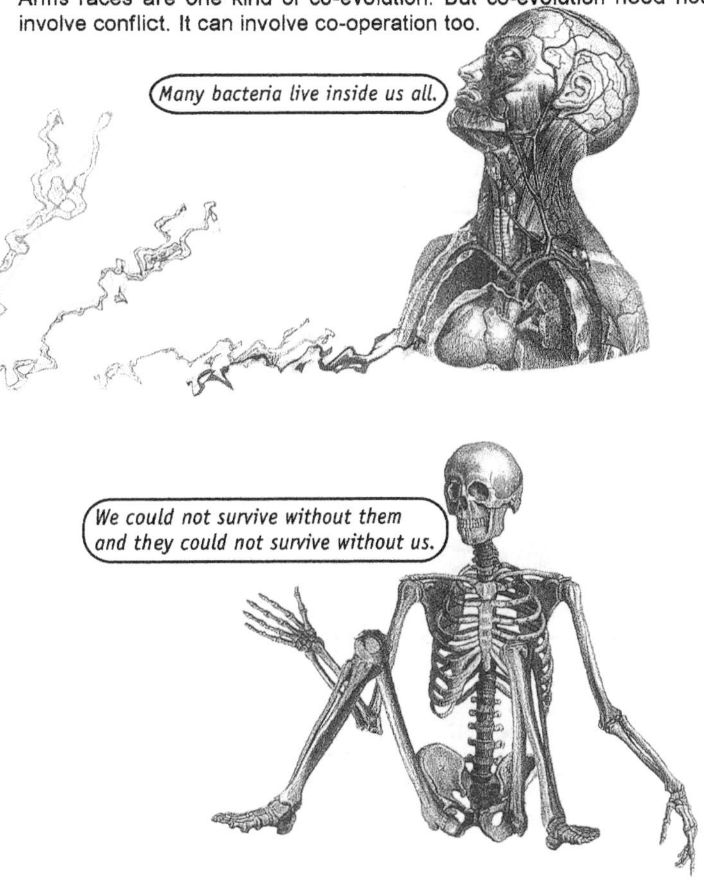

They have co-evolved with us in a co-operative relationship that biologists call "mutualism".

The Origin of the Eukaryotes

Sometimes, when two species co-evolve in a co-operative way, they can become so interdependent that they actually fuse into one species. This is probably what happened when the eukaryotes first evolved some 1.8 million years ago.

Eukaryotes are one of the two "empires" into which biologists classify all living things on earth. The other empire is known as the prokaryotes.

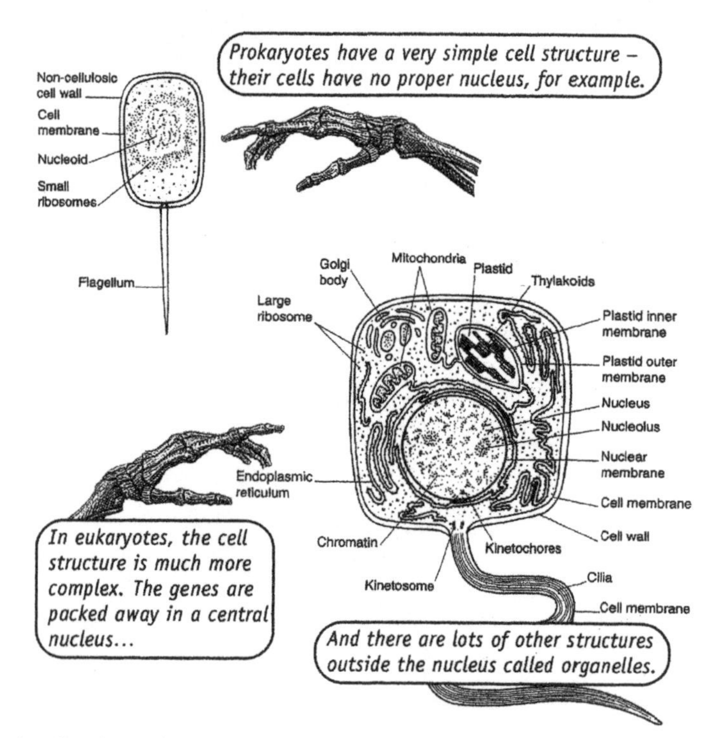

In all eukaryotic cells, there are little structures called mitochondria. These provide the energy for the rest of the cell. Most biologists today believe that the mitochondria were once free-living prokaryotes that gave up their autonomy and allowed themselves to become part of another organism.

Other Differences

Prokaryotes and eukaryotes differ in many other ways besides their cell structure. One such difference is size. All prokaryotes are single-celled organisms. Some eukaryotes are single-celled organisms too.

Another difference is that all prokaryotes are asexual, while many eukaryotes reproduce sexually.

Let's Talk About Sex

When biologists talk about sex, they refer to a particular way of having offspring. All organisms make copies of themselves (have offspring), but not all of them do it sexually.

Virgin Births

In asexual reproduction, one parent gives rise to a new cell which then develops into a new organism on its own, without the need for fertilization. This is called *parthenogenesis*, or "virgin birth".

Clones and Genetic Difference

In species that reproduce without sex, each offspring is *genetically identical* to its single parent (barring the occasional mutation). In other words, each offspring is a *clone* of its parent.

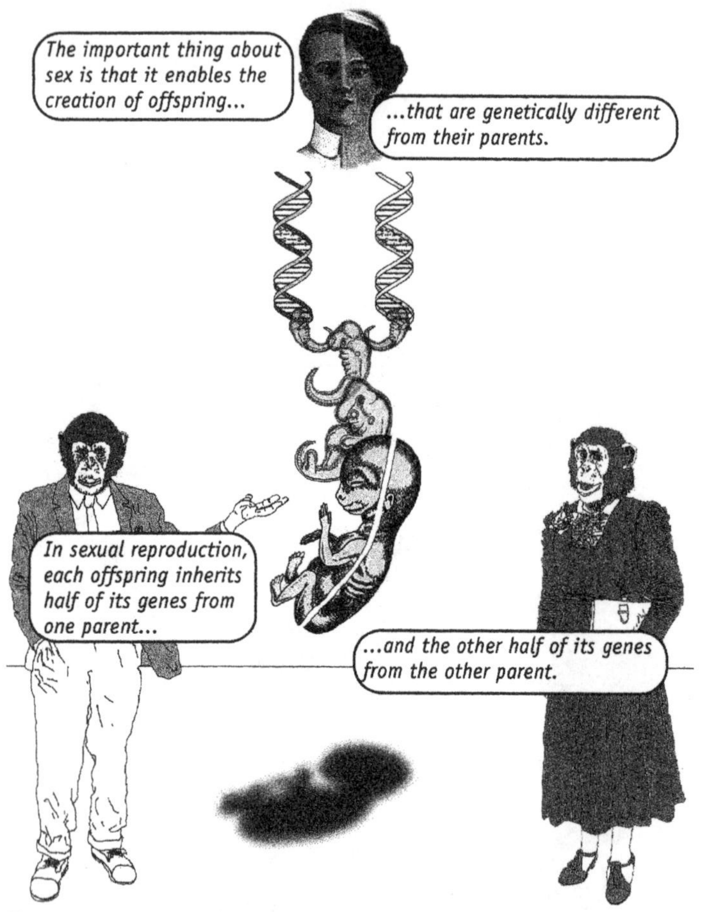

The result is an organism with a completely new combination of genes, unlike either parent.

Meiosis and Gametes

In most sexual species today, the two cells (one from each parent) that fuse together to create the zygote are different from the other cells in the parents. Most cells in each parent contain two sets of genes. When cells divide by mitosis in the process of development, each new cell usually receives a copy of each set of genes.

However, not all cell division is mitosis.

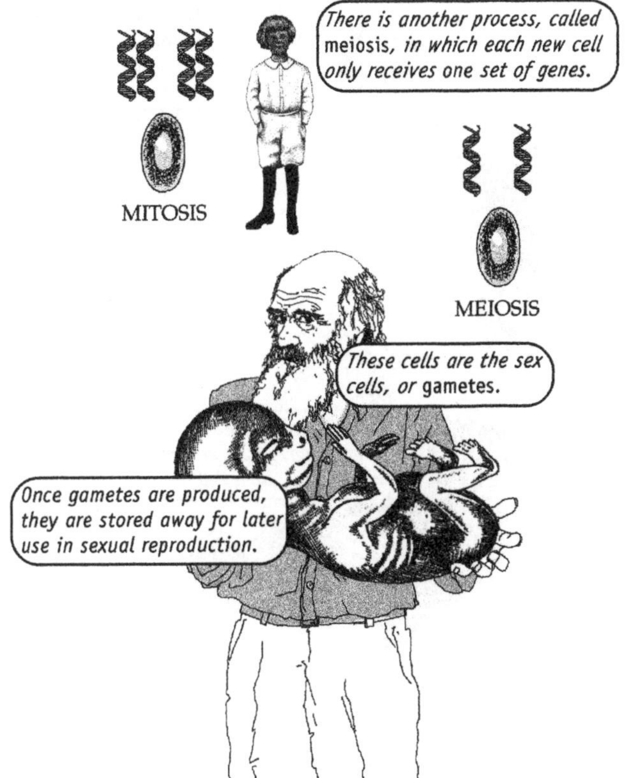

At the moment of fertilization, two gametes from separate organisms fuse to create a new cell – a zygote – which becomes the first cell in the new organism. The zygote has two sets of genes, one from each parent.

Male and Female

In sexual reproduction, there are always two parents, but there are not always two sexes. The evolution of two distinct sexes – male and female – came later than the evolution of sexual reproduction.

For biologists, the essential difference between the two sexes lies in the gametes (the sex cells that fuse to form a zygote).

All the other things that we usually associate with sexual differentiation – such as penises, vaginas, breasts and beards – are ultimately just consequences of the fundamental difference between sperm and eggs.

Isogamy and Anisogamy

The first sexually reproducing organisms were *isogamous*, meaning that they all produced gametes of only one size.

Many single-celled organisms are still isogamous. But most multicellular organisms are *anisogamous*, meaning that there are two sexes which produce different types of gamete.

The Cost of Sex

The first organisms to evolve were all asexual. But now most species are sexual. This seems to indicate that sex must convey some advantage.

When you reproduce *asexually*, all the genes in your offspring come from you. But when you reproduce *sexually*, only half the genes in your offspring come from you. So, sex halves the chances of any gene getting into the next generation! Why would a selfish gene put up with this?

What's the Advantage of Sex?

Biologists still disagree about the answer to this question. One theory is that sex allows harmful mutations to be dispensed with quickly. In an asexual species, when a harmful mutation occurs in one individual, all its offspring inherit it. Gradually, harmful mutations accumulate over the generations, and may eventually cause that lineage to die out.

So, the good genes accept their lowered chances of being passed on to the next generation in return for the increased chance of being passed on with other good genes.

The Family Swap

Another theory is that sex allows beneficial mutations to accumulate more quickly. In an asexual species, for beneficial mutations to end up in the same organism, the mutations have to occur in the same family, one after the other.

In a sexual species, however, beneficial mutations can end up in the same organism even if they first occur in different families.

Reproductive Isolation

Sometimes, a sperm meets an egg but cannot fuse with it. When this happens, no offspring is conceived. The male whose sperm it was, and the female whose egg it was, do not become parents.

The two populations can be thought of as having completely separate "gene pools", because they cannot swap genes with each other.

Gene Pools

The gene pool is the set of all the genes in a sexually reproducing population at a given time. It is called a pool because genes that are found in separate individuals today can (because of sex) potentially be found together in the same individual in the future. Each time a new individual is born, it is as if a bucket is dipped into the gene pool and a new combination of genes is selected at random.

Genes found in separate asexual individuals will never be found together in the same individual in the future, *unless sex evolves*.

The Biological Species Concept

According to one definition, known as the "biological species concept", a species is a set of organisms that can interbreed with each other. On this definition, when two populations are reproductively isolated (when they have separate gene pools), they constitute different species.

Like the concept of the gene pool, the "biological species concept" only works for species that reproduce sexually. Asexual organisms never "interbreed", so we must find some other definition of species that applies to them. There is not just one way to define what a species is.

The Phenetic Species Concept

According to another definition, known as the "phenetic species concept", a species is a set of organisms that resemble one another and are distinct from other sets. There are many problems with this definition, since the notion of *resemblance* is so vague.

After all, the usual way in which we recognize what species an organism belongs to is by looking at it. If we see that a particular bird has a big bright tail with two iridescent eye-like designs on it, for example, we know that it is a peacock and not a robin.

Speciation

Speciation is the process by which new species come into being. What counts as a "new species" depends, of course, on how you define what a species is.

According to the biological species concept...

Like the biological species concept on which it is based, this way of defining speciation only applies to sexual species.

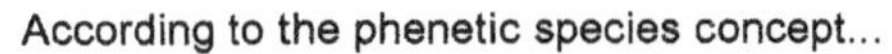

According to the phenetic species concept...

Splitting

Common to both definitions of speciation is the notion of splitting. New species always arise when a single species splits into two daughter species. The only difference between the two definitions of speciation lies in the criteria that we use to tell one species apart from another.

When discussing speciation, we should always make it clear which definition of species we are using.

As an example, we will take the biological species concept and show how evolution can lead one population of interbreeding organisms to split into two daughter populations that are reproductively isolated.

One way this can happen is by geographical separation. Suppose that a colony of rats lives happily in a large valley. They can all interbreed, so they are all members of the same species.

For hundreds of years, the two populations are geographically separated, and each evolves in slightly different ways.

Allopatric Speciation

Then, one day, the water drains away and the two populations meet again. By now they are so different that they cannot interbreed.

According to the biological species concept, this means that the two populations now constitute distinct species. Where once there was one species, now there are two. When speciation occurs by geographical separation, as in this story, it is known as **allopatric speciation**.

Higher Groups and Lower Groups

The species is the most important way of grouping organisms. For purposes of classification, biologists also recognize higher and lower groups. Below the species level, biologists distinguish different varieties, or "subspecies". Different varieties all belong to the same species because they can interbreed, but have some distinguishing features.

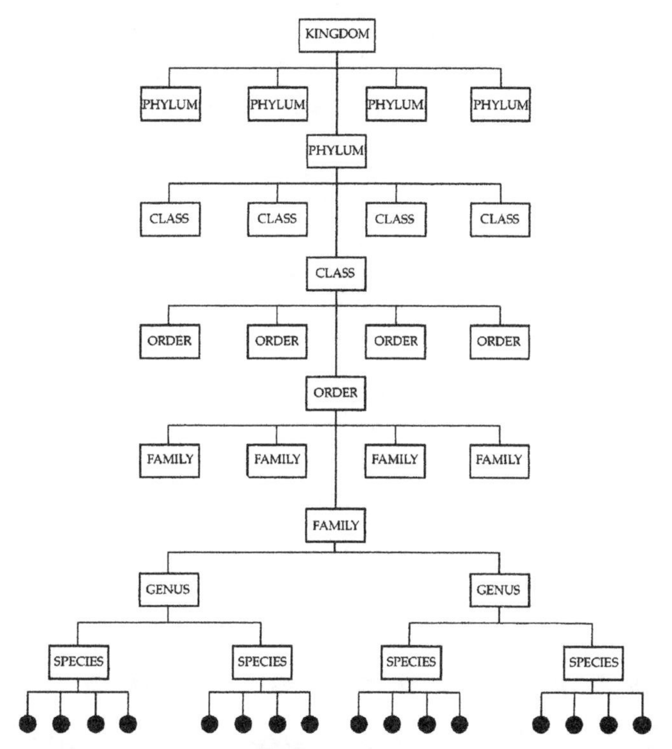

Above the species level, biologists recognize a number of higher groups. First, different species are grouped together in the same *genus* (plural: *genera*). Then several different genera are grouped together in the same *family*. Several families combine to make up an *order*. Orders combine to make a *class*. Several classes make up a *phylum*, and several phyla create a *kingdom*.

Varieties of Dogs

For example, the varieties of domestic dog are all (just about) members of the same species, *Canis canis*. The second term here tells us the *species*; and the first is the name of the *genus*. The genus *Canis* also contains other species such as the grey wolf (*Canis lupus*) and the golden jackal (*Canis aureus*).

The genus *Canis* combines with several other genera, such as the fox genus (*Vulpes*), to make up the family *Canidae*. *Canidae* belongs to the order *Carnivora*, which belongs to the class *Mammalia*, which is part of the phylum *Chordata*, which is part of the kingdom *Animalia*.

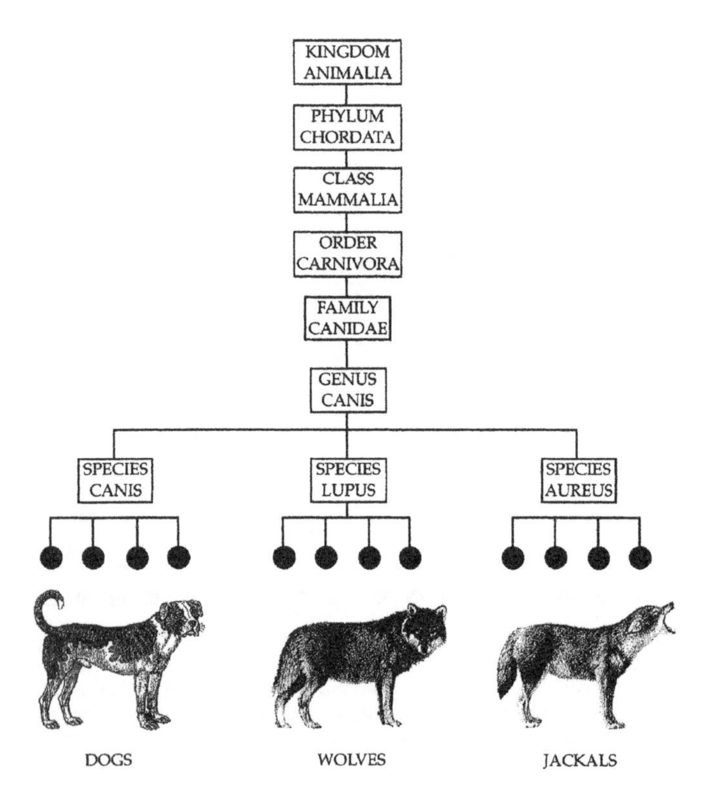

Empires and Domains

What group comes above the kingdom? What is the highest group? It used to be thought that the highest group was the *empire*. As we have already seen, there are two empires: the prokaryotes and the eukaryotes (see pages 94 and 95).

EMPIRES

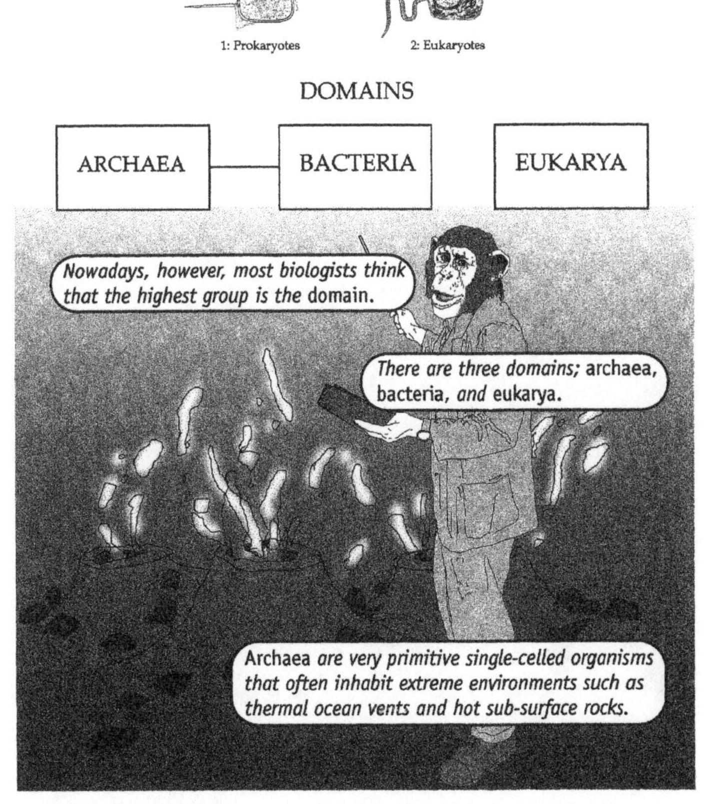

There is not much difference between these two answers, since archaea and bacteria are both types of prokaryote. The only difference lies in the order in which the domains evolved.

Which is the Highest Group?

To say that the highest group is the *empire* is to make an implicit claim that the first speciation event in the evolution of life on earth involved the eukaryotes branching off from the prokaryotes. Only later did the prokaryotes split into two groups: the archaea and the bacteria.

Those who argue that the highest group is the domain object to this view. They claim that the tree of life is rather different. Some genetic evidence suggests that the first split was between the bacteria and the other two groups. Only later did the eukaryotes branch off from the archaea.

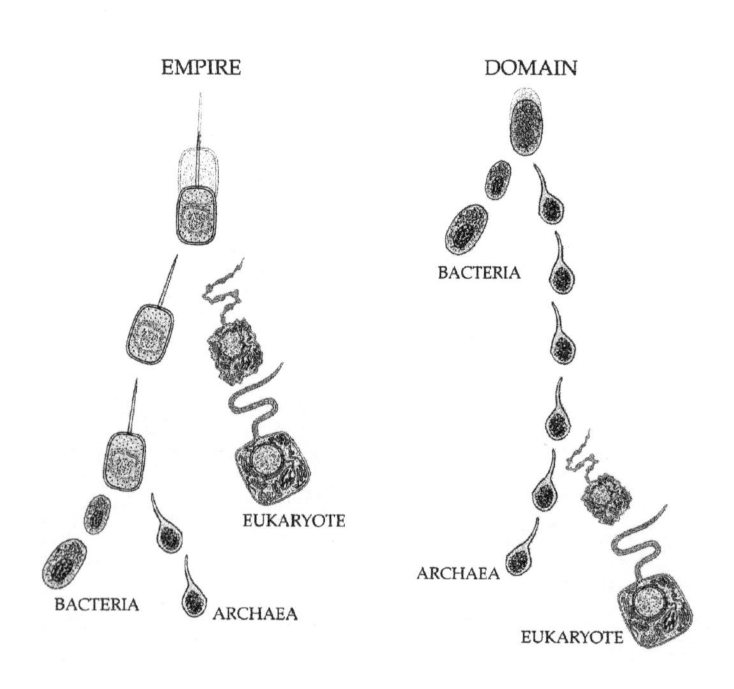

Monophyletic and Paraphyletic Groups

Another way of describing the difference between these two views is as follows. Dividing all living things into two empires implies that the prokaryotes are a *monophyletic* group, while dividing organisms into three domains allows that prokaryotes might be a *paraphyletic* group.

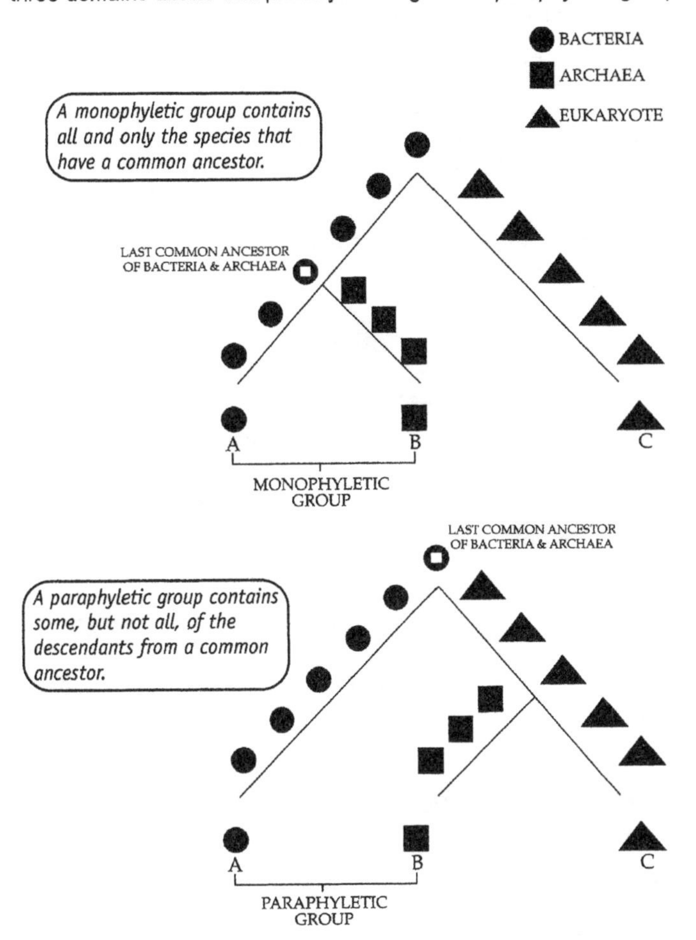

Cladism

According to a school of thought known as *cladism*, biologists should not use paraphyletic groups for classifying organisms because they obscure the evolutionary relationships between them.

If we follow this rule, biologists would have to abandon some old ways of classifying organisms. For example, they would have to give up thinking about fish as forming a single group.

Mammals and other four-limbed creatures (tetrapods) are all descended from one particular group of fish, the lobe-finned fish, as the diagram below illustrates. You can see from the diagram that the category "fish" is a paraphyletic group. It includes lobe-finned fish and ray-finned fish, but not the tetrapods. So, according to the cladistic approach, biologists should either redefine fish or not really use the term "fish" in their classification systems.

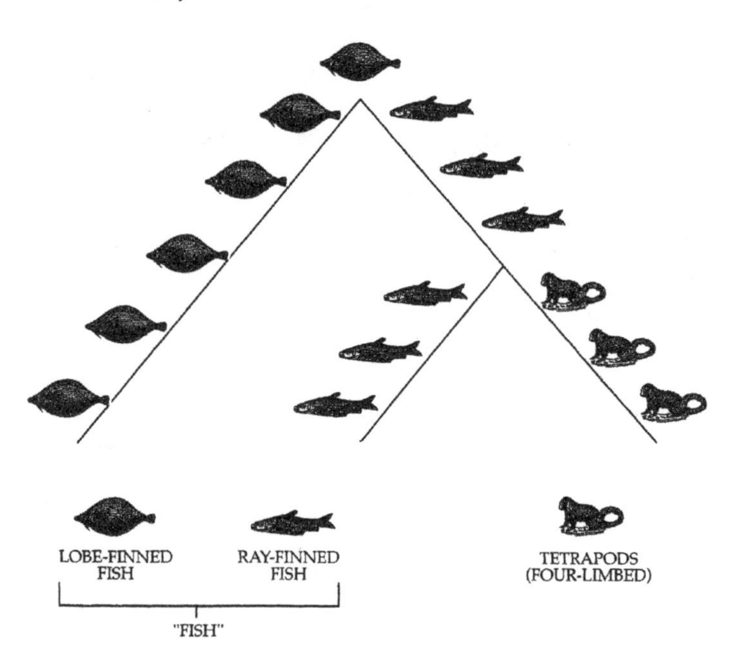

The Tree of Life Again

In the cladistic approach, the hierarchical classification system of species, genera, families, orders, classes, phyla, kingdoms and empires/ domains reflects the history of evolution.

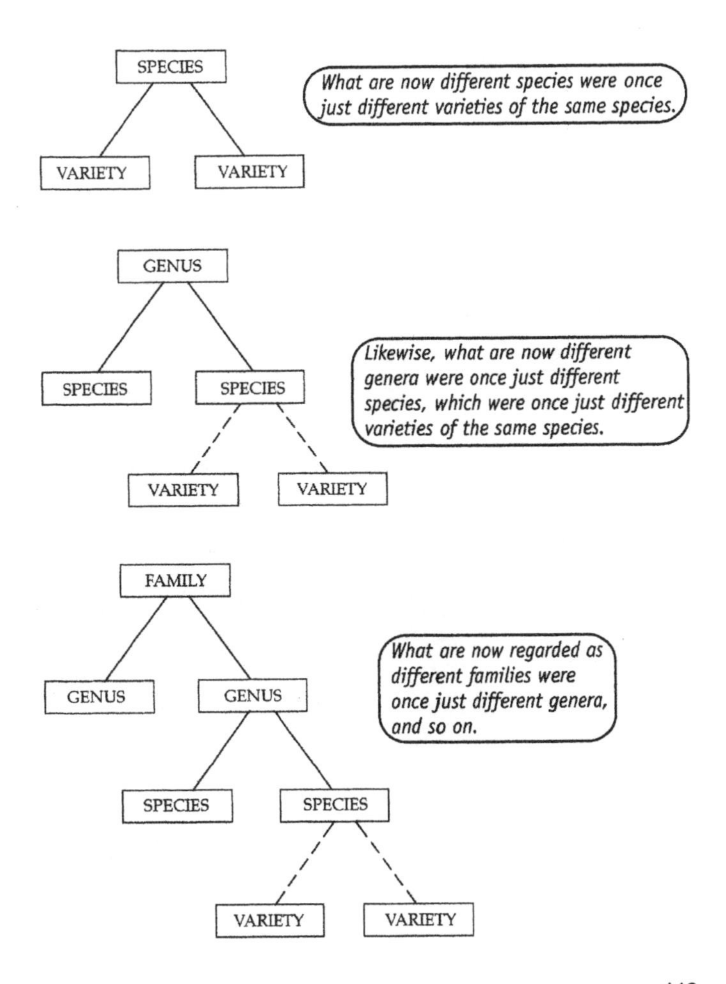

The Trunk of the Tree

We can carry on this argument right up to the highest group, the empire or domain. What we now regard as two empires or three domains were, a very long time ago, just different varieties of the same species. At that time, near the beginning of life on earth, all living things on this planet were members of the same species. We can visualize this as the trunk of the tree of life.

The millions of different species that are alive today are all little twigs on the tree of life, ultimately descended from one single ancestral species.

The Origin of Life

But where did the first species come from? How did life on earth get started in the first place?

Darwin himself said very little about this question.

The first living thing cannot have come into existence from another living thing. It must have sprung from *non-living* matter. How could this happen?

A Warm Little Pond

Some years after *The Origin of Species* was published, Darwin did offer a suggestion about the origin of life. In a letter written in 1871, he speculated on how life might have first appeared...

Building on Darwin's suggestion, the Russian biochemist **Aleksandr Oparin** (1894–1980) and the Scottish geneticist **J.B.S. Haldane** (1892–1964) proposed, in the 1920s, that energy from the ultraviolet sunlight and from lightning discharges might have synthesized organic compounds as it passed through the primitive atmosphere.

The Primordial Soup

Rather than Darwin's "little pond", however, Haldane thought that these compounds would be washed into the sea.

In 1953, the American chemist **Stanley Miller** (b. 1930) tested this idea by passing an electric current through a chamber containing water, methane and ammonia. Just as predicted, a wide range of organic compounds was formed. These organic compounds could have come together to form the first life forms.

Near the Sea-bed

These ideas suggest that the first life forms may have arisen near the surface of the sea. More recently, however, some biologists have suggested that life first emerged far *below* the surface, near the hydrothermal vents found at the sea-bed. Hydrothermal vents are places where underwater volcanic activity produces plumes of high-temperature water which rise from the sea-bed.

These organisms are called *hyperthermophiles*, because they require high temperatures to survive. Perhaps the first life forms were hyperthermophiles.

Panspermia

Another possibility is that the first living things on earth originated from outer space. Perhaps life first arose on some other planet, and traces of it were carried to earth on, say, a meteorite. This idea is called *panspermia*, which means "seeds everywhere".

Panspermia is not as far-fetched as it sounds. It may sound like science fiction, but those who investigate the origin of life on earth consider it to be a real possibility.

There is even hope that primitive bacteria may be found on Europa, one of the moons of Jupiter. And there is increasing evidence that life once existed on Mars. Scientists have found what looks like traces of microbial activity on the red planet. But if life ever did flourish on Mars, it died out a long time ago.

Whether life on earth started here, or came from outer space, biologists would still like to know *how* life got started in the first place.

Extinction

Evolution is not just about the birth of new species. It is also about the death of old species, or "extinction". A species can go extinct for many different reasons.

Mass Extinctions

During most of the history of life on earth, the number of species has tended on average to increase slightly each year. Every year, a few new species come into existence, and a few old species go extinct. But the number of new species is usually slightly more than the number of species going extinct.

For a while, thousands of species go extinct each year. The total number of species is drastically reduced. These moments are called "mass extinctions".

The End of the Dinosaurs

There have been at least five really huge mass extinctions in the history of life on earth. The most recent one happened about 65 million years ago.

These mighty beasts, who had ruled the earth for millions of years, left no descendants, except perhaps for the birds. Whenever you see a chicken, or a sparrow, remember that its ancestor might have been a dinosaur.

Why Did the Dinosaurs Go Extinct?

What caused the mass extinction 65 million years ago in which the dinosaurs and many other life forms died out?

In 1980, the physicist **Luis Alvarez** (1911–88) suggested that this extinction was caused by the impact of a large asteroid. There was only one problem with this theory. A large asteroid impact would leave a massive crater on the earth's surface.

Through the 1980s, the missing crater posed a major problem for Alvarez's theory. Then geologists discovered a massive crater buried beneath sediments on the Yucatán coast of Mexico. The size and date of the Chicxulub crater are consistent with Alvarez's theory.

A Hundred Million Megatons of TNT

How could an asteroid impact have caused a mass extinction? A massive asteroid impact would have produced an explosion equivalent to that produced by a hundred million megatons of TNT. Such an explosion would have thrown up a global dust cloud, which would have blocked out sunlight for several years until the dust settled.

Like a row of dominoes toppling over, there would have been a cascade of disasters rippling through the food chain. The impact of a large asteroid could also have triggered other disasters, such as global warming, acid rain, volcanic activity and massive forest fires.

The Lucky Ones

The mass extinction did not kill everything, however. Some species managed to survive. Among the survivors was a small, insignificant group of animals that looked rather like tree-shrews. They were the ancestors of all the mammals alive today, including humans.

While the dinosaurs had ruled the earth, the early mammals had lived in the margins.

If that asteroid had not crashed into the earth 65 million years ago, the dinosaurs might well be around today, and the mammals might still be lurking around in the shadows. Without that asteroid impact, the human race might never have evolved.

The Sixth Mass Extinction

Mass extinctions are not just a thing of the past; there is one going on right now. For the past few decades, thousands of species have been going extinct each year. The diversity of life forms is being drastically reduced.

Unlike the previous five, however, the cause of the latest mass extinction is not an asteroid or some climatic upheaval. It is due to the action of one species – **us**.

Ecological Disaster

Human beings are wreaking havoc on the natural world. The great rainforests of South America and South-East Asia are dwindling at an alarming rate. The greenhouse effect is causing the world to heat up and the polar ice to melt. Towns and cities are increasing in number and size, eating into the countryside around them. Chemical and nuclear wastes pollute the rivers and seas.

The expansion of the human race, and its amazing technological progress, spells disaster for the natural habitats of thousands of species. Whole ecosystems are in danger of collapse.

The End of the Human Race

The last mass extinction wiped out the dinosaurs and cleared the way for the mammals to inherit the earth. Who will be the victims of the current mass extinction? And who will be the survivors?

There would be some poetic justice in this. We would have inadvertently driven ourselves to extinction. Some have suggested that insects will survive the current mass extinction, even if humans don't. Will cockroaches inherit the earth?

Is the Earth One Giant Organism?

Perhaps nothing will survive the current mass extinction. Perhaps humans will manage to wreak so much havoc on the natural world that all life on earth will end.

Some people like to think that the earth will not let this happen.

The view that the planet is like one huge organism is a nice idea. It's also wrong.

Organisms Are Designed By Natural Selection

There is a simple reason why the planet cannot be regarded as a single huge organism: organisms are self-replicating entities designed by natural selection.

Natural selection requires competition between slightly different self-replicating entities.

Therefore, the earth cannot be regarded as an organism. Such comparisons are misleading and based on nothing but wishful thinking. Unfortunately, there is no reason why life on earth must continue forever, or even for much longer.

Gaia

The view that the earth is a single giant organism is sometimes referred to as the "Gaia hypothesis". This term (from the Greek concept of "Mother Earth") was first introduced into modern biology by the English scholar **James Lovelock** (b. 1919) to refer to a rather different idea.

Lovelock argued that there were some very interesting feedback mechanisms linking biological and climatic processes. These feedback mechanisms could link living things and atmospheric events into a single system with a limited capacity for self-regulation.

However, even the scientists who do not agree with Lovelock accept that there is a world of difference between his scientific hypotheses and the crackpot romantic myth of the earth as a giant organism which defends itself against parasitic human beings by any means necessary.

The Deadly Species

Some people blame the current ecological disaster on new technology. They urge us to go back to a time when humans lived in harmony with nature. Unfortunately, this too is romantic nonsense. Humans have always been a deadly species.

Before humans entered Europe, Asia and the Americas, there were dozens of species of large land mammal roaming wild. Now there are very few.

The only continent where big game can still be found roaming wild is Africa, where animal populations have had much longer to evolve ways of avoiding human predators. New technology may have helped humans to destroy the natural world *more quickly*, but that is all. Our ancestors still managed to do a lot of damage without it.

Killing Our Cousins

Among the many species that the human race has helped to drive extinct were various species of hominid. The hominids first appeared around three million years ago. Unlike the other primates, hominids walked on two feet rather than four, and had a very human-like appearance.

As with the other mammals that our ancestors encountered in their various migrations, the non-human hominid species may well have been the victims of human success.

The Neanderthals

The last hominid species to go extinct was *Homo neanderthalensis*. The Neanderthals probably left Africa much earlier than our ancestors, perhaps around 300,000 years ago. By the time our ancestors arrived in Europe, some 50,000 years ago, the Neanderthals had already been there for quite some time.

The last Neanderthals probably lived in Spain some 35,000 years ago.

The Advantage of Language

The Neanderthals were very similar to us. Their brains were as big as ours, perhaps even bigger. But for some reason they never developed any lasting material culture. And they lacked the one thing that gives humans a massive advantage over every other species on the planet: **language**.

Some think it was as long ago as 250,000 years. Others think it was much more recent, perhaps only 100,000 years ago. One thing is clear though. Language enabled humans to conquer the planet.

Efficient Information-sharing

Language allows information to be shared among the group much more effectively than other forms of communication. Other primates use their vocal chords to convey information.

Extended Memory

But language is not just a communication system. It also allows you to think in completely new ways.

Texts are like external memory that can be plugged into your brain, vastly increasing your brain's computing power.

The Speaking Animal

Some people think that the human capacity for language means that we can completely override our evolved instincts. Postmodernists, for example, often write as if language were everything. According to this extreme view, humans exist in a cultural world that has completely broken free from the natural one. In some ways, this is reminiscent of the views of the English philosopher **John Locke** (1632–1704), who claimed that the human mind is like a "blank slate" on which different cultures can "write" anything they want.

If it weren't, how could we explain the fact that there are so many similarities between the various cultures of the world?

Human Universals

Cultural diversity dazzles and impresses us, but we shouldn't let it obscure the important fact that people all over the world have many things in common. All cultures have the following things: language, myth, dance, gesture, distinct sex roles, levels of social status, sexual regulations and greater public dominance by men.

Some universals might have been "invented" once, when all humans lived in a single cultural group in Africa, and then retained by all the different groups that branched off from the original ancestral population. Others probably spring more or less directly from **human nature**.

What is Human Nature?

The fact that all the various cultures from around the world have so many things in common suggests that the human minds that create these cultures also share many characteristics, and this, in turn, suggests a common evolutionary history. It seems that the human mind has been shaped by natural selection just as much as the human body.

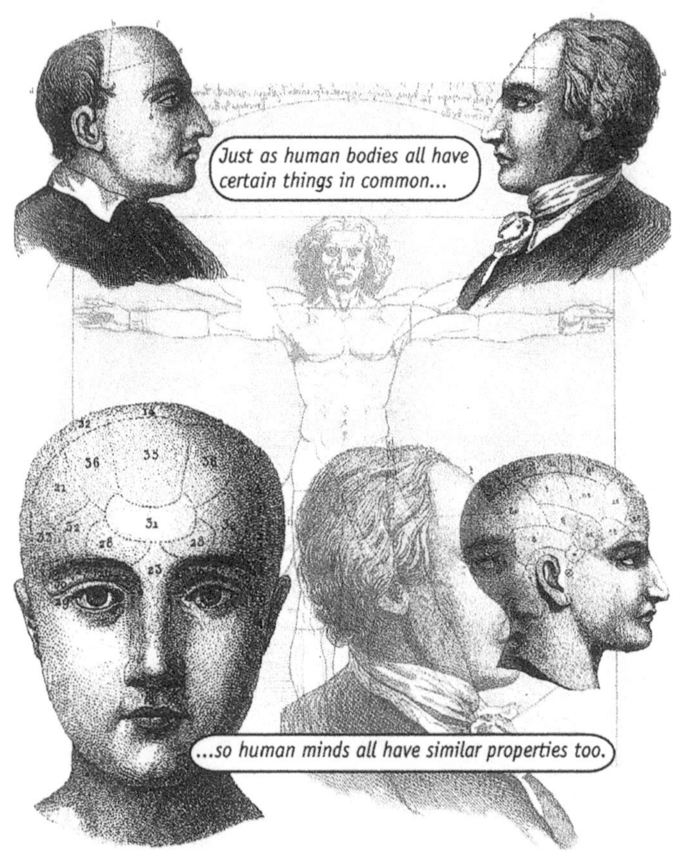

The universal design features of the human mind constitute "human nature". They are the legacy of thousands of years of evolution.

Evolutionary Psychology

Evolutionary psychology is the study of the human mind from an evolutionary perspective. It is the scientific study of human nature.

In the words of **Leda Cosmides** and **John Tooby**, the mind is not a general-purpose problem-solver, but a collection of lots of particular problem-solvers. It is more like a Swiss-Army knife than a single blade.

Massive Modularity

The particular problem-solvers in the human mind are called **modules**. Scientists have found some evidence that our capacities for vision and language are subserved by modules. But some evolutionary psychologists go much further. Cosmides and Tooby argue that much more complex capacities, such as reasoning about social situations, are also subserved by modules.

Not all evolutionary psychologists accept the massive modularity hypothesis. Some believe that there are modules for things like vision and language, but not for reasoning in general. In this vein, the archaeologist **Steven Mithen** (b. 1960) has suggested that the distinctive thing about the evolution of the *human* mind is that the barriers between different bits of the mind were eliminated.

The Walls Came Tumbling Down

According to Mithen, the pre-human mind was structured a bit like a medieval cathedral. Just as these cathedrals had a central nave surrounded by a number of small chapels, so the pre-human mind had a central area of general intelligence, surrounded by some specific modules for things like language and tool-making.

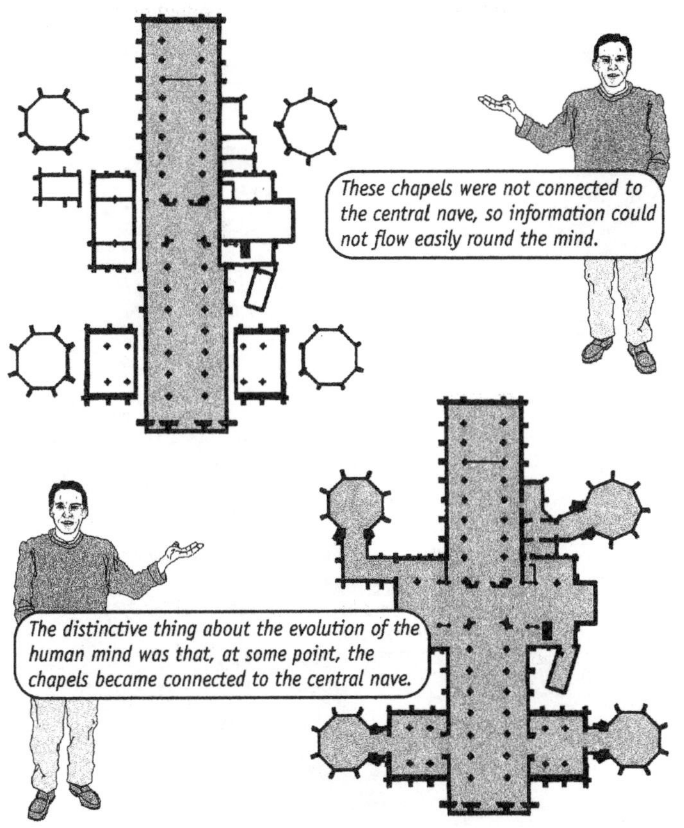

The information that was previously trapped in distinct modules could now flow freely around the mind, giving rise to the unique cognitive flexibility that only humans seem to have.

Science and Speculation

Mithen's ideas, like many of those put forward by evolutionary psychologists, are very speculative. There is nothing wrong with speculation, however. All scientific theories begin as speculation. The important thing is that they don't remain that way. Once a theory is formulated, it must be tested. If it passes tests, it becomes accepted knowledge. If it fails the tests, it is corrected or abandoned.

We can compare humans with less similar species such as chimpanzees, but this doesn't tell us much about the uniquely human things, such as art and morality.

Survival Machine or Courtship Device?

When evolutionary psychologists do get round to examining the uniquely human things like art and morality, they usually explain them as by-products of mental capacities that evolved for more practical purposes. For example, in his book *How the Mind Works*, **Steven Pinker** focuses on pragmatic capacities such as vision and child-rearing...

In Miller's view, all the most distinctively human capacities are courtship devices.

Is Art a Sexual Display?

According to Miller, the things we think of as most "human", such as our capacities for art, morality and language, did not evolve because they provided any survival benefit, but rather because they helped our ancestors to seduce their lovers.

In suggesting that our artistic capacities function as sexual adverts, Miller is not suggesting that all art is motivated by some unconscious wish to have sex.

All that is needed for Miller's theory to work is for our ancestors to have preferred to mate with artistic partners.

Humans and Culture

Miller's theory is controversial and awaits rigorous testing. The evolutionary basis of our artistic capacities is still an open question. Whatever the answer may be, it is undeniable that the capacity for culture is more highly developed in humans than in any other species.

There are, for example, songs that are specific to local populations of some bird species, and nut-cracking techniques that are specific to local groups of West African chimpanzees. These cultural traditions, however, are very limited compared to the rich variety of human culture.

The Evolution of Culture

Some people have suggested that human culture evolves in the same way that genes do. In 1976, Richard Dawkins coined the term "meme" to denote the hypothetical unit of cultural evolution.

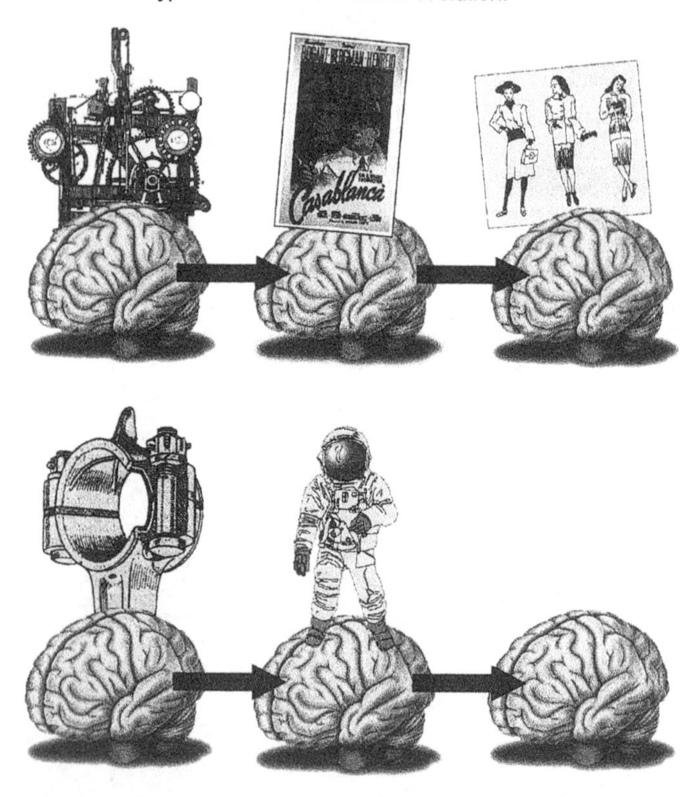

"Examples of memes are tunes, ideas, catch-phrases, clothes fashions, ways of making pots or of building arches. Just as genes propagate themselves in the gene pool by leaping from body to body via sperms or eggs, so memes propagate themselves in the meme pool by leaping from brain to brain via a process which, in the broad sense, can be called imitation." (Dawkins, *The Selfish Gene*, p. 192)

Memes or Mind-viruses?

The French cognitive scientist, **Dan Sperber** (b. 1942), agrees that culture can be said to evolve. Unlike Dawkins, however, Sperber thinks that the units of cultural evolution are more like viruses than genes.

Fertile ideas are like parasites that use our brains as vehicles for their propagation. According to Sperber, the study of cultural transmission should be more like *epidemiology* (the study of how diseases spread) than genetics.

Why Do Some Ideas Catch On More Than Others?

Whether ideas are more like genes or viruses is still an open question, but both these theories agree that culture evolves as ideas become either more or less popular.

Evolutionary psychologists argue that ideas become more successful when they fit well with the evolved properties of the human mind.

The Naturalness of Religious Ideas

One of the most successful ideas in the history of humanity has been the idea of a divine being. Gods and other supernatural entities have figured in virtually every culture. Why does this idea fit so well with the human mind?

The French anthropologist, **Pascal Boyer**, has argued that religious ideas are natural in many ways.

It is a short step from here to the idea that the sun rises, or rain falls, because some *supernatural* person wants it to. The concept of a supernatural being also grabs our attention because it is both familiar and strange. Gods are supposed to be like people in some ways but not in all.

Evolutionary Epistemology

According to some philosophers, scientific theories also evolve. And just as it is only the fittest organisms who survive, so only the most accurate theories prosper in the scientific community. This account of scientific knowledge is known as "evolutionary epistemology".

As we have just seen, religious ideas are among the most successful ideas in the history of humanity. So, it seems that even the most inaccurate ideas can prosper. Ideas become popular when they fit with the evolved propensities of the human mind, and this criterion does not always coincide with that of accuracy.

Evolutionary Ethics

What about ethics? Do ideas about right and wrong evolve too? And, if they do, what makes some ethical ideas prosper while others die out?

Evolutionary psychologists argue that moral codes become popular for the same reasons that other kinds of ideas prosper – because they dovetail with evolved features of the human mind.

Moral Evolution

The idea that humans have moral instincts may strike some people as very un-Darwinian, but that is because, to many, Darwinism is all about survival of the fittest, and so leads to a view of humans as selfish animals.

However, we have already seen how selfish genes can prosper by building altruistic organisms.

Moral Instincts

When faced with moral dilemmas, people often act unselfishly.

In one experiment, for example, hundreds of wallets, containing a few dollars and a fake ID, were left lying around the streets of New York City.

There are many other examples of such instinctive altruism. In many countries, people give blood for no reward. And in fact, in the US where people are offered cash for donating blood, donors are rarer than in the UK, where they receive no financial compensation.

Moral Sentiments

From the cases of instinctive altruism, it seems that people are often guided not by rational self-interest, as many economists think, but by emotions that incline them to help others, such as sympathy.

This was what **Adam Smith** (1723–90) thought.

His first book, *A Theory of Moral Sentiments* (1759), was published a hundred years before *The Origin of Species*, but it foreshadows much recent evolutionary thinking in its view of humans as guided by ethical instincts.

Are Ideas Alive?

For some thinkers, to say that ideas *evolve* is just a metaphor. For others, it is literally true. Some go even further, claiming that this means that ideas are actually living things. Richard Dawkins has argued that the capacity to evolve by natural selection is in fact the defining feature of life.

If we ever discover life on another planet, we can be sure that it will be the product of an evolutionary process.

Artificial Life

In the 1980s, researchers in artificial intelligence began to use Darwinian ideas to design new computer programs. Instead of writing the programs themselves, these researchers generated a few initial programs randomly. They then allowed these programs to compete in solving a problem. The programs that performed best were allowed to make copies of themselves, but the copying process was made deliberately imperfect, so that some errors would occasionally be made. Then the new programs were allowed to compete again, and so on.

That is why this field of artificial intelligence is called "artificial life".

Genetic Algorithms

Computer programs that evolve in this way are called "genetic algorithms". Genetic algorithms are becoming more important as they are used for an increasing number of purposes, from managing investment portfolios to designing bridges and buildings.

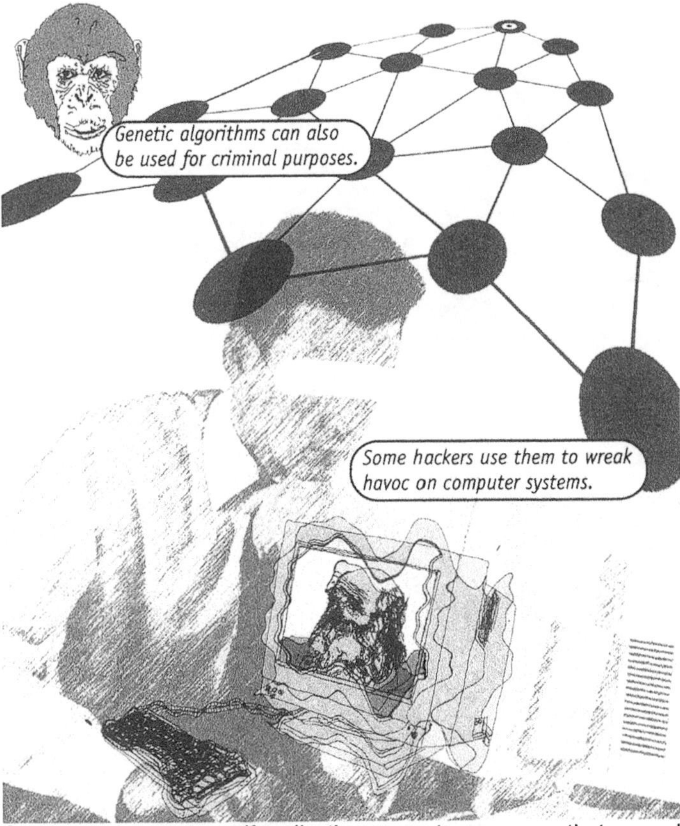

Computer viruses are self-replicating computer programs that spread from one hard disk to another by means of floppy disks and e-mails. Just like real viruses, these artificial parasites can cause immense damage.

Evolutionary Robotics

Genetic algorithms are also being used to design robots. This area of artificial intelligence is known as "evolutionary robotics". In 2000, two American scientists at Brandeis University, called **Hod Lipson** and **Jordan Pollack,** used genetic algorithms to allow robots to design and build themselves.

The robots in Lipson and Pollack's experiment were very simple things consisting only of a few articulated bars. However, there is no reason why the same techniques could not be used, in the future, to evolve more complex robots. Perhaps, one day, robots will evolve to be more intelligent than we are.

The Virtual World, *Tierra*

Experiments in evolutionary robotics are expensive and difficult to run. For the time being, therefore, most experiments in artificial life involve simulations of evolution inside the computer rather than in the real world.

A classic example of this approach is *Tierra*. *Tierra* is a virtual world, designed by **Thomas Ray** in the early 1990s. After designing *Tierra*, Ray populated it with a simple genetic algorithm, and then sat back and watched.

Different species of digital organism evolved, and competed with each other for space on the computer's hard disk. Ray was observing a case of genuine evolution *in silico*.

A Common Thread

These discussions of artificial life and evolutionary robotics may seem far removed from the biological questions with which we began this book. But they are all linked by Darwin's simple yet powerful idea. The theory of evolution by natural selection is not only at the heart of modern biology, but is becoming increasingly important in many other fields, from artificial intelligence and psychology to philosophy, anthropology and even sociology.

And unlike other influential ideas from the 19th century, such as Marxism and Freud's psychoanalysis, the theory of evolution has not been discredited. On the contrary, the evidence in its favour is greater than ever.

The Controversy Continues

Despite the overwhelming evidence in its favour, the theory of evolution by natural selection is still not widely accepted or understood. As we saw at the beginning of this book, even in the United States, one of the most technologically advanced countries in the world, Darwin's ideas are still the subject of fierce controversy.

In Kansas, the story has a happy ending. In February 2001, just a year and a half after the theory of evolution was removed from the school curriculum there, the decision was reversed. Thanks to the efforts of a state-wide network of scientists, the central idea in biology was once again taught in Kansas classrooms.

The battle between science and superstition is not over yet...

Surveys show that between a quarter and a half of Americans still believe in the literal biblical account of creation. It seems that many people prefer to ignore the scientific evidence rather than abandon or modify the religious explanations they were taught as children.

A Scientific Answer to the Deep Questions

The creation myths of Christianity and other religions provide answers to the deep questions: *Is there a meaning to life? Who are we? Why are we here? What is man?*

After posing the last of these questions, the biologist **George Gaylord Simpson** declared...

Since Darwin gave us the theory of evolution by natural selection, we have had a genuinely scientific answer to the deep questions. No longer do we have to resort to superstition to answer them.

The scientific answers to the deep questions have struck some people as dull and uninspiring compared with the colourful myths of religion. Even the English biologist **Thomas Henry Huxley** (1825–95), who did so much to defend Darwin's theories, remarked that scientific progress was tragic.

The Wonder of Evolution

To Darwin, however, the fact of evolution was not ugly at all. The final sentence of *The Origin of Species* expresses his wonder at the beauty of our evolutionary history.

"There is grandeur in this view of life, with its several powers, having been originally breathed into a few forms or into one; and that, whilst this planet has gone cycling on according to the fixed laws of gravity, from so simple a beginning endless forms most beautiful and most wonderful have been, and are being, evolved."

Science and Values

This is not to say that science can tell us everything. Science can't tell us how we should live, for example. Science can inform us only about how things *are*, not about how things *should* be.

The fact that humans have evolved from apes does not force any particular moral conclusion. Our evolutionary past is one thing. Our future is quite another.

Further Reading

Darwin is his own best introduction. Nobody should go through life without reading *The Origin of Species*, or at the very least the first five chapters. Its simple elegant prose is as moving and as clear today as it was when the book was first published in 1859. Many editions are available, including one in the Penguin Classics series.

For more recent introductions to evolutionary theory, I recommend the following two books:

The Blind Watchmaker, by Richard Dawkins (UK: Penguin, 1988; US: Norton, 1988). Dawkins is unsurpassed in his talent for explaining Darwinian theory to a lay audience. This book gives a vivid account of how natural selection works, and corrects many common misunderstandings.

Darwin's Dangerous Idea: Evolution and the Meanings of Life, by Daniel C. Dennett (UK: Penguin, 1995; US: Simon & Schuster, 1995). In addition to giving a very clear explanation of how evolution by natural selection works in biology, this book also explores the consequences of Darwin's idea for broader issues in psychology, ethics and philosophy.

If you want something more technical, you might like to try the following:

Evolution, by Mark Ridley (get the second edition: Blackwell, 1996). This mammoth textbook, with over 700 pages, provides a comprehensive technical introduction to evolutionary theory. Intended for undergraduate biology students, it can also be used for self-study.

Evolutionary Genetics, by John Maynard Smith (get the second edition: Oxford University Press, 1997). This is another undergraduate textbook that can also be read by lay readers who are willing to put in some effort. More mathematical than the book by Ridley, it provides a solid introduction to the aspects of genetics that are vital to understanding evolution.

If you are interested in the problem of altruism and/or the theory of sexual selection, the following two books are a must:

The Selfish Gene, by Richard Dawkins (get the second edition: Oxford University Press, 1989). First published in 1976, this book is a brilliant introduction to the work of Bill Hamilton, Robert Trivers and the other evolutionary biologists who solved the problem of biological altruism.

The Ant and the Peacock, by Helena Cronin (Cambridge University Press, 1992). This book gives a very clear account of how evolutionary biologists solved the problem of biological altruism (the ant) and the puzzle of sexual selection (the peacock).

For an overview of the whole history of life on earth, the following books are recommended:

The Variety of Life: A Survey and a Celebration of all the Creatures that have Ever Lived, by Colin Tudge (Oxford University Press, 2000). This clearly written and well-illustrated book covers the whole tree of life, explaining the evolutionary relationships between all terrestrial organisms, past and present. A monumental achievement by one of the best science writers around.

The Origins of Life, by John Maynard Smith and Eörs Szathmáry (Oxford University Press, 1999). Two prominent biologists trace the history of life on earth by describing what they call "the nine major transitions in evolution".

If you are particularly interested in human evolution, try the following three books on evolutionary psychology:

Introducing Evolutionary Psychology, by Dylan Evans and Oscar Zarate (UK: Icon, 1999; US: Totem, 1999). What can I say? Another cartoon book by me! Oscar's illustrations are wonderful.

How the Mind Works, by Steven Pinker (UK: Penguin, 1998; US: Norton, 1997). A much more detailed introduction to evolutionary psychology by one of the pioneers of the discipline. A joy to read.

The Mating Mind, by Geoffrey Miller (UK: Heinemann, 2000; US: Doubleday, 2000). Miller presents his alternative version of evolutionary psychology, according to which many aspects of the human mind evolved through sexual selection.

The Authors

Dylan Evans is a Research Fellow in the Department of Philosophy at King's College London, where he is involved in a project on the evolution of the emotions. He is the author of *Emotion: The Science of Sentiment* (Oxford University Press, 2001) and (with Oscar Zarate) *Introducing Evolutionary Psychology* (UK: Icon, 1999; US: Totem, 1999).

Howard Selina was born in Leeds, and studied painting at St Martin's School of Art and the Royal Academy. He works in London and Yorkshire as a painter, illustrator (and aspiring songwriter), and is currently distracted from these and other tasks by trying to renovate an old steel narrow-beam cruiser. This is his third book for Icon.

Acknowledgements

The author would like to thank Oliver Curry for reading a previous draft of the manuscript and providing valuable constructive criticism. I am also grateful to all the other members of the Darwin@LSE Work-in-Progress Group for helping me to deepen my understanding of evolutionary theory during the past few years. Thanks are also due to Richard Appignanesi and Jennifer Rigby of Icon Books for their editorial support and eagle-eyed attention to detail. Last but not least, I am grateful to Howard Selina for numerous suggestions and helping to make this book such fun to write.

The artist would like to thank Ms Judy Groves for letting him have his philosophers reference book back, Dylan Evans for his help with the diagrams, and Ms Paola di Giancroce for the cups of tea and the roll-ups.

Index